COMPUTER INTERFACING
connection to the real world

Martin Cripps

*Director of the Wolfson Microprocessor Unit and Senior Lecturer,
Department of Computing Imperial College, London*

Edward Arnold
A division of Hodder & Stoughton
LONDON NEW YORK MELBOURNE AUCKLAND

© 1989 Martin Cripps

First published in Great Britain 1989

For Clare, Stephanie and Andrew

Distributed in the USA by Routledge, Chapman and Hall, Inc.
29 West 35th Street, New York, NY 10001

British Library Cataloguing in Publication Data
Cripps, Martin
   Computer interfacing: connection to the real world.
   1. Computer systems. Interfaces
   I. Title
   004.6

   ISBN 0-7131-3588-3

Typeset in 10/12pt Times by Keyset Composition, Colchester
Printed and bound in Great Britain for Edward
Arnold, the educational, academic and medical publishing division of
Hodder and Stoughton Limited, 41 Bedford Square, London
WC1B 3DQ, Page Brothers (Norwich) Ltd

Preface

- Real-time and on-line control and logging systems are among the most difficult of computer systems to implement

- More systems fail because of inadequacies in their interfacing and input-output handling than from any other cause.

- The explosive increase in the availability of cheap computer power has meant that a large proportion of such systems are based on microprocessors.

- Almost all microprocessors have inadequate or even faulty input-output structures and represent a step backwards from many earlier minicomputer designs.

- Research time, design effort and expenditure on input-output has been paltry by comparison with that put to other aspects of computer architecture and languages.

The five points listed above are the reason for this book. They represent a situation which cannot be allowed to continue. There are no reasons except history and inertia which prevent the universal adoption of correct, well-structured interfaces and the input-output structures to support them. There are even signs that language designers are realizing that 20% of a system takes 80% of the effort and that most of that difficult 20% is handling input-output: **the connection to the real world!**

Complex systems of the type described here are generally a collection of different kinds of hardware and software, often using unrelated technologies which must work as a 'total system'. Interfacing these diverse parts together with the outside world has always been difficult due to a lack of standardization and often a lack of understanding of the underlying requirements. While we may be unlikely to find a single interface system philosophy to solve every design problem, a small set of standard interfaces and approaches should suffice in all but the most extreme cases.

This book attempts to provide the fundamentals of theory and practice necessary to achieve this. It is concerned with the structure and design of interfaces, particularly those connecting to the real world as this is more difficult than simply linking to computer-like peripherals. The use of programmable logic and its techniques are emphasized in the design of interfaces and existing standard interfaces are described with their shortcomings.

The book starts from the real world and its variables and continues by considering how to transfer these variables into the computer in a useful and accurate form. Transducers and sensors, some sampling theory, signal conditioning and conversion, are all concerned with getting the values we need from the wide, analog world into the narrow, digital, binary form needed in a computer. By linking well-founded theory to practical application the firm underlying structure of real-world interfaces is introduced. A later section on environmental constraints covers the side effects of running a system which may undesirably affect the environment, and those effects from the outside world which may cause malfunctions in the system. Again firmly based theory allows for straightforward practice.

The sections on standard interfaces show where the structured approach first loses momentum. For mainly historical reasons most standards are a bit of a mess. Even some quite recent ones make life more difficult than it need be. The general section on interfacing shows how this could be corrected.

Almost all interfaces today are either to microprocessors or use microprocessors in their implementation. There are many fundamental principles involved in the design of micro-processors but unfortunately the majority of those sold so far do not subscribe to them.

The architectures necessary to support good high-level languages and well-organized input/output are discussed. Only recently have integrated circuits become dense enough to hold the necessary transistors, and their designers able enough to apply them. Specialized integrated circuits to support input-output are considered in detail followed by design techniques and, most importantly, testing of implementations to ensure that they are reliable. The book ends with some examples and an extensive bibliography for further reading.

The book was developed to support advanced courses in computer interfacing at Master's and final-year undergraduate levels. Interfacing, like input-output in general, is still a Cinderella area compared with the effort expended on programming and other parts of system design. It is hoped that industrial and home computer users wishing to widen their horizons beyond programming will also find this a worthwhile introduction.

I wish to thank my students over many years for their comments, mostly polite, which have helped to refine the text. In particular, my thanks go to my colleague Tony Field who provided many valuable improvements. I would also like to thank Jane Spurr for her patience in the preparation of the typescript.

Martin Cripps
1987

Contents

1

Introduction

Figure 1.1 shows the problem domain of interfacing and the scope of this book. Though modern computing commenced in the early 1950s, all consideration of interfacing both between units and, more importantly, to the outside world was extremely *ad hoc* until the late 1960s. To simplify the overall hardware of early systems, almost any complexity was tolerated in connections between units. The cost of this approach was poor, one-off interface designs, and we still have to live with the legacy in many awkward standards established at that time.

A revolution in interfacing techniques has been brought about by the availability of cheap programmable logic in the form of microprocessors, but early examples gave rise to as many problems as solutions. The first microprocessor (the Intel 4004) was introduced at the start of the 'seventies and is already of only historical interest. It was scarcely more powerful than a simple pocket calculator and was quickly followed by the first of the eight-bit processors the 8008 and then by a rash of others. All of these early microprocessors suffered from poor input/output structure, inadequate architectures for sophisticated programming support, and poor interfacing. This was due to the limitations of the number of transistors which could be fabricated on a silicon chip, the size of a chip, and the complexity with which designers could cope. Many of the inherent limitations of those microprocessors, and even some of the designs themselves, are still with us because of commercial pressures.

The modern approach to interfacing, as with high-level languages, is to get the specification correct and then (and only then) to design an architecture to support it. This puts the complexity of connections into the hardware, but current large-scale integrated circuits can easily provide a far greater level of complexity than is needed for well-designed interfaces.

The first microprocessor with an adequate architecture for this purpose was the Texas Instruments Inc. 9900 and was largely based on their earlier 960 range of minicomputers. TI managed to pack a full sixteen-bit processor on a chip no larger than many inadequate eight-bit designs. This was achieved by using a 'workspace' off the chip rather than having specific registers on it. This approach supports multi-process operation easily and is gaining in popularity in the latest designs. Similarly by removing specialized I/O instructions the Motorola corporation saved chip area in their range of microprocessors and provided a much cleaner interface between processing and external data. This has, in turn, led to vastly improved interface design.

1.1 Structured Interfacing to the Real World

We should perhaps really start by considering the type of system with which we are concerned. Computers are traditionally associated with computation, though more recently with office automation, but a far wider field of application exists. This involves connection to the real world. This may be for data gathering or for on-line control but the complexity which is added by dealing with irrational behaviour is considerable. If a computer is used for

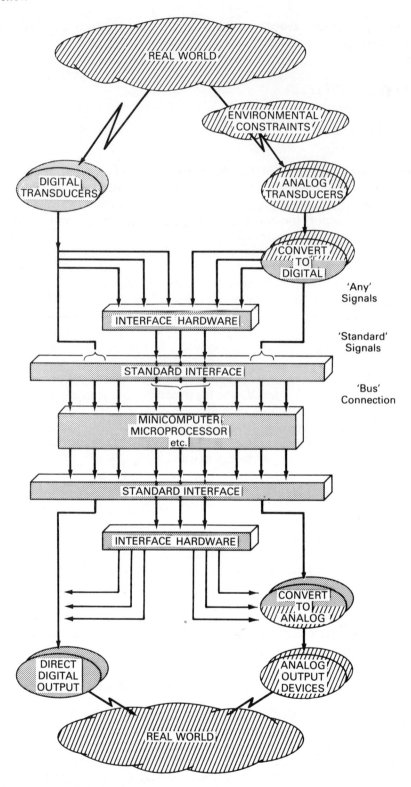

Fig. 1.1 Interfacing to the real world

accounting, planning, word processing, etc, then the inputs to it are well behaved: key strokes and digital values. The real world on the other hand will be analog and often unpredictable due to noise. If a fault occurs in a chemical plant, for example, then the computer will have to go from its normal steady controlling function to panic actions which will load or even overload it.

These systems which connect to the real world can be split into three types.

- **Data gathering, logging and monitoring** systems are those which only have real-world inputs. All outputs are to traditional computer peripherals such as displays or printers.

- **Stimulator and annunciator** systems are those which only have real-world outputs, taking their inputs from traditional peripherals like keyboards.

- **On-line control** systems, on the other hand, have both inputs from, and outputs to, the real world in addition to any traditional peripherals. As they represent the most interesting type we can split them up further by three factors:

  - Is the control **supervisory** or **directly digital**?
  - Is the control operated by an **open** or a **closed** loop?
  - And finally, is the system required to operate in **real-time** or not?

To expand on these points, **supervisory control** uses a computer interfaced to conventional analog instrumentation and control electronics which is itself responsible for controlling the given process. The traditional analog control system uses electronic circuits which monitor a given variable and provide outputs to drive suitable valves, heaters, etc. to provide the control. The variable may be monitored directly (**proportionally**), or it may be integrated over a period to allow its past history to affect the control, or it may be differentiated to derive its rate of change. The analog control circuits in most common use were combinations of two terms (**pi** or **pd**) or three term controllers (**pid**). To include supervisory control a computer reads data from the analog signals (after conversion to digital form) and uses this to adjust the 'set-points' of the analog controllers. The computer also records data and provides mimic displays of it to the plant operators in a form they would expect. This style of control was the easiest to add as it left the original two- and three-term controllers in place and required minimal retraining of operators. The large UK chemical company Imperial Chemical Industries (ICI) designed a system called Media which can be operated in this way and is in widespread use in their plants.

More recently the computer, now a number of microprocessors, has been used to replace the analog controllers completely. The raw variables are input from transducers directly and compared with stored values to produce the linear (proportional), integral and differential terms to generate the control outputs. Operator displays of all variables and their history are available, usually on visual display screens rather than analog meters. This style of **direct digital control** was first tried in the early 'sixties but the increased reliability of mini- and microcomputers is what has now made it quite feasible to remove all the analog controllers.

Closed loop operation means that the computer system is fully responsible for the control actions and has feedback paths to affect them. In some industries such control is not acceptable because we will not trust machines and insist on a human link in the chain. An example of such an **open loop** system is a passenger aircraft where, despite all the computer aids, we insist on a pilot completing the control loop. Requiring that the system operates in **real-time** simply places minima on the computer power and data rates that are required to affect the control. Of course, one can say that every system has deadlines, the ones **we** put upon it. But it is when the constraints of the real world impose strict time limits that 'real' real-time operation becomes essential.

One of the many improvements in overall systems design is taking place due to the low cost and high reliability of microprocessors. In factories and laboratories the various monitoring and control points may be widely spaced. Early direct control systems linked each point individually to a central computer via its associated interface hardware. This gave rise to cable congestion, problems with reliability, lack of flexibility in adding new points, and to the use of the 20 milliamp current loop interface. Surprisingly there is still no international standard accepted for this common industrial interface! With perhaps a thousand points spread over hundreds of metres such systems were expensive to construct even using multiplexers to share lines back to the centre. The use of a common bus for interfacing to local mini- or microcomputers, and then the use of standard data communication links to the centre gives a distributed computer system of great flexibility, which may, in many ways, be considered simpler than the centralized approach.

This design approach can be taken much further. As pre-amplifier chips are very cheap the early stages of the interface hardware can be mounted with the transducer. With modern production technology this would mean integrating a transducer and pre-amplifier on the same substrate or chip.

One can take this approach much, much further. As the cheapest microprocessor chips cost very little more than a pre-amplifier, one can include **all** stages of the interface, signal gathering and conditioning with the transducer. The output of the combined unit is then a standard serial or network interface. Thousands of transducers and their connection cables from the system described before then become thousands of microprocessors linked by a standard network, and yet the total system is simpler to understand and operate.

1.2 A Micro as the Interface or an Interface to the Micro?

We now have to be slightly schizophrenic in our approach. Figure 1.2 shows the problem. Do we consider a microprocessor and its accoutrements to be a computer or to be a component?

The term 'transputer' was coined by Iann Barron to describe a single chip computer including processor, store and interfaces that could be used as a component, just like a transistor, in a larger circuit (hence *trans* from transistor and *puter* from computer). At that time the combination of 'transducer and computer' alluded to earlier would have been just as strong a candidate for the name and was as important a breakthrough. That name would imply a computational transducer or an information acquirer and processor. There is a discussion of the Barron transputer approach to computing in Chapter 13.

In a wide variety of interfaces we can replace discrete, random logic by a microprocessor or microcomputer. The only places when such a replacement is not practical are those where the hardware must operate at very high speed or would have to perform too many operations in parallel. In circumstances other than these extreme ones, a microprocessor

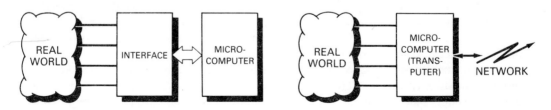

Fig. 1.2 Interface to the micro and the micro as the interface

becomes just **part of** the interface and its system design is approached in that light. The complexity of integrated circuits has increased to the point where we can include all the stages of analog circuitry, which will be discussed in following chapters, with the digital processing elements. Such a chip located with a transducer becomes the entire interface up to a single standard input-output.

The alternative approach is to consider the interface as a way of connecting **to** a microprocessor which is then to carry out the control or data logging, etc. Common personal computers are often used for experimental control and data logging purposes. To ease the load on the processor caused by getting data in and out, a second (third, fourth, . . .) processor is easily justified. An obvious example is the International Business Machines Inc. range of personal computers. They have a small microprocessor simply to interface the keyboard to the main 'processing' processor. So throughout this book we will consider what parts of an interface can be replaced by a micro and, at the same time, what we need to consider in an interface from the real world to a microprocessor.

1.3 The Microprocessor or the Microcomputer?

Microprocessors were developed almost by accident from their calculator origins. It took only ten years from the discovery of the transistor in 1947 to the first, albeit simple, integrated circuit. A further 14 years saw the introduction of the first microprocessor in 1971 with *only* 2300 transistors on the chip. It was a four-bit processor because of this size limitation. A further seven years and 70,000 transistors on a chip gave a full 16-bit microprocessor with few architectural restrictions. Seven more years and we could have half a million transistors on a chip. A full computer with processor, store and input output is now easily practical. So some difficulty with terminology has arisen. What is a microprocessor? Figure 1.3 shows a range of processors compared for basic system cost and their power and architectural quality. They run from simple four-bit machines suitable for use in children's toys, to those machines with more power than traditional minicomputers. Some of the commonly used names for chips and boxed-up computers of these types are listed below.

Chips	Boards and Boxes
Microprocessor	Programmable Logic Controller (PLC)
Single Chip (Micro)Computer	Single Board Computer (SBC)
Signal processor chip	Microcomputer
Transputer	Personal Computer (PC)
Bit-slice processor	Minicomputer

Any computer consists of a processor, or processors in some cases, read-only and read-write storage, and input-output. The names indicate the nature of integration of the elements. A **microprocessor** is a single chip implementation of the central processor of the chosen architecture. It includes the arithmetic and logic unit, register bank, instruction register and sequence controller, bus controller and bus buffer drivers. The pin interface contains the address, data and control signals and power lines.

A **single chip micro** computer contains not only the processor but also both types of store and input-output circuits. The pins can contain just those for I/O but may be arranged to give a normal bus to access more store if needed. There are a number of varieties of single chip design, some giving quite dramatic versatility. The first is to replace the read-only store, programmed by masking at the time of manufacture, with erasable (reprogrammable)

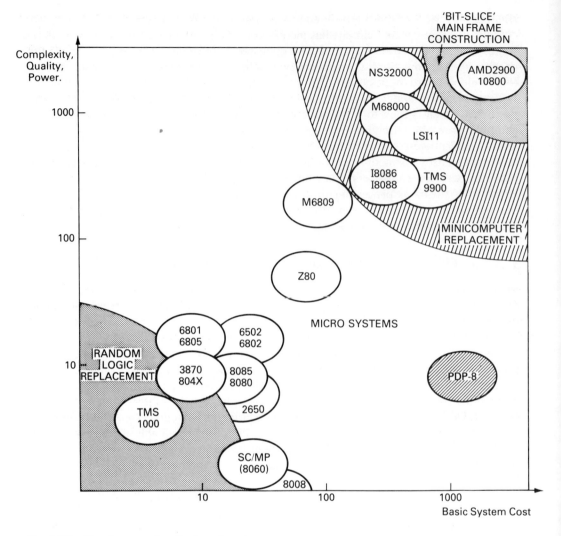

Fig. 1.3 Cost/performance/applications for micros

read-only store. These chips can be used in small numbers, there being no need to have thousands with the same program to make them cost effective. The I/O of the single chip microcomputer is usually digital, though some provide basic analog functions. The digital side commonly provides a mix of many individual inputs and outputs which can be handled in parallel, one or a few serial communication lines, and special interface circuits which permit inputs or outputs to be counted or timed at speeds faster than processor instruction sequences could manage.

Single chip microcomputers are available which include some of the analog circuits to be discussed in later chapters. Analog values can be input directly, be converted to digital form and then be processed. They can then be used to control digital outputs or be reconverted and output in analog form. A special breed of chip includes a very fast hardware multiplier circuit. This takes up quite a large area of the chip but makes for very fast running of specific

signal processing algorithms. The final single chip variety is the **transputer** which has already been mentioned. A very high-speed processor (sometimes with hardware for floating-point operations) and store are connected to serial data links operating at 10 million bits per second, or more. These allow the transputer to be used individually or linked in arrays as just a component of more powerful systems. The final variant gives an alternative approach to producing high-power computer systems which is to use **bit-slice** processors. The system designer can build up different architectures from chips each of which contains a slice of the arithmetic and logic unit (ALU), registers, stack, etc.

The various board or boxed machines have in addition to any of the chips above a power supply, more store and a variety of peripheral devices. For many areas of on-line control, particularly in factories, the **programmable logic controller** has been the dominant technology. The original relay devices have been replaced by microprocessor-based versions, but they perform the same functions. They are programmed to give reliable but simple sequences of outputs depending on the states of inputs from sensors. The word **logic** is sometimes dropped from the name so that we can confuse this PC with the personal computer. A **personal computer** usually sits on your desk and includes a full visual display screen and typewriter-like keyboard along with floppy and sometimes hard discs. It is used for wordprocessing, financial spreadsheet, database or similar applications. They are also widely used as a base for computer aided design workstations. I regret that I cannot give a definition for the term minicomputer as it is (ab)used so widely that I no longer know what it means! If you spent between 20 and 200 thousand dollars then you probably bought a mini!

1.4 Sampled Data Systems

As soon as we ask our microprocessor to do a job for us we meet a problem. Let's say it has some inputs, some processing requirement and some outputs. Figure 1.4 shows an automobile application. A set of inputs from the engine, the environment around it and the driver's right foot are taken into the microprocessor. It then computes the optimum time for ignition and drives an output to fire the spark. The problem is that the system is not continuous in operation. It must only **sample** its inputs, and then compute, and then output. It is unlikely that the system will spend more than 10% of its time on the inputs. In the

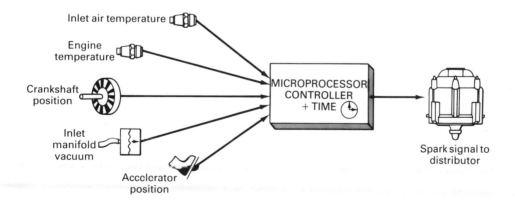

Fig. 1.4 Typical 'sampled data' microprocessor system

example there are five inputs, so that is 2% for each. That means the system does not look at the inputs for 90% of the time. Now imagine driving your car with your eyes shut for 90% of the time. Do you think the result would be completely accurate? I don't advise you to try it. The following chapters investigate the problem for computers which must operate in this fashion, and define solutions.

2

Input Transducers

In the real world variables are temperature, pressure etc., but our computer can only cope with encoded binary (digital) numbers which represent the variables. The first step in the chain from any real world variable to a number in the computer is called a **transducer**, or occasionally a sensor.

The definitions of what constitutes a transducer are many, varied and for the most part confusing. For example Thompson's

> *'any device in which energy is changed from one form to another, irrespective of whether there is amplification and irrespective of the mode of presentation of the result'.*

This would include the conversion of the sun's energy into growth of a plant, which would not fit most users' idea of a transducer. As there is no satisfactory definition as yet, a statement of the purpose is probably of more use.

> A transducer used with a computer system converts information (signals) which are available in one form to another form which, for input, is electrical in nature and can be acquired and subsequently processed. An input transducer responds to a physical stimulus and produces an electrical signal with a known relationship to it.

One serious snag which always has to be borne in mind is that a transducer is the one part of a system which **has to be trusted** as it is not possible to self-test except in a few cases. As it is the ultimate connection to the variable, only by replication can there by any check on its correct operation.

There are so many types of transducers that an entire book would be needed to do them justice but the principles can be extracted and the more important examples will be used to demonstrate them. Transducers can be categorized by three main headings.

(a) **By the variable being measured**
(b) **By the measurement technique of the transducer**
(c) **By the transducer type**

The first category, the variable being measured, is by far the most important division, the others simply serving to identify the limits on usage. The other important principle, apart from the way in which the transducer operates (how it works) is its **accuracy**. This determines its usefulness in practice.

2.1 Input Variables, Classification and Units

Input variables can be grouped by the units in which they are measured. The SI (System Internationale) provides the most widely used system of measurement units and divides them into three types.

(a) **Basic Unit quantities (MKSAKC)**
(b) **Derived Unit quantities (combinations of basic units)**
(c) **Dimensionless quantities (ratios and hence unitless)**

The six basic measures are Length (Metre), Mass (Kilogram), Time (Second), Electric Current (Ampere), Thermodynamic Temperature (degree Kelvin) and Luminous Intensity (Candela). Common derived units are Force (Newton = $kg\, ms^{-2}$), Pressure (Bar = 10^5 Newtons per square metre) and velocity (metres per second). More than 70% of all transducers in current use are used for sensing temperature, pressure, displacement and flow.

A typical dimensionless quality is chemical content, usually expressed as a percentage or for smaller quantities as parts per million (ppm). Some variables are expressed in dimensionless form as no absolute can be reliably found. An example is sound; what is absolute quiet? Table 2.1 lists all the commonly transduced variables and their units of measurement with some ways of converting the variable to an electrical signal.

The electrical signal output will vary from one transducer to another and will depend on the power supply for those transducers which need a supply. There have been attempts at standardization, none totally satisfactory, of which the 20 mA loop is most common, others being ±10 V, 1–5 V, 5 mA and the 4–20 mA derivative of the 20 mA loop. It is assumed at present that there will be no standard until 'intelligent' transducers are common and a computer data transfer interface is used.

It can be seen from Table 2.1 that there is a wide variety of choices of transducer for many variables, only a few being listed, and the categorization by transducer type and measurement technique helps to show up some limitations.

The types are **Active** and **Passive** with the distinction being that the former require no external power source to operate, such as a flow meter formed of a paddle wheel incorporating a fixed magnet which passes a coil inducing a current in it, whereas passive transducers require an external current source, such as a potentiometer to measure rotational position. The distinctions of measurement technique are rather more important than this as most transducers are passive anyway.

Measurement Technique

(a) **Direct**   (a) **Digital**   (a) **Incremental**
(b) **Indirect**   (b) **Analog**   (b) **Absolute**
(c) **System**

Direct transducers convert from the physical stimulus to an electrical signal. Indirect transducers require an additional stage where the input variable is converted to an intermediate form which is then converted to the electrical signal related to the original variable. Figure 2.1 shows both types being used to measure liquid level in a tank. The direct transducer uses a capacitor, the value of which is altered by the liquid acting as a dielectric over part of the area. This gives an electrical signal, from the passive sensor, proportional to the level. The capacitance increases as the tank fills. The indirect transducer, on the right, has a float which moves with the liquid thus converting the level to a rotational movement. This displacement is then measured using a potentiometer. The significance of the intermediate stage is that it can give rise to more error as there are two conversion stages and a link between them.

Table 2.1 Input variables and units

Variable	Units	Techniques
Acceleration	$m\,s^{-2}$	Displacement and Time, Force on Spinning Gyroscope.
Composition	%, ppm	Very varied but includes change in conductivity or semiconductor device operation due to absorption at surface.
Density	$kg\,m^{-2}$	Weight (by strain of gauge) of known volume, differential pressure.
Displacement	metre	Optical grating, resistive or capacitive change due to movement of parts, microswitch.
Electric current Electric potential	ampere volts $= kg\,m^2\,A^{-1}\,s^{-3}$	Direct amplifier or attenuator.
Energy	joule $= N\,m$	Domestic wattmeter, heat energy by calorimeter.
Flow (rate)	$m^3\,s^{-1}$	Cross-section plus vane, turbine, laser scatter, doppler ultrasonic, differential pressure.
Force	newton $= kg\,m\,s^{-2}$	Stress–strain, strain gauge.
Frequency	hertz $= s^{-1}$	Counter (or event counter—by proximity capacitance, light etc.).
Humidity	% (relative)	Conductivity change of exposed surface (dewpoint) photoelectronic.
Luminous intensity	candela	Photo-diode, transistor, resistor.
Magnetic flux	weber $= Vs$	Changes: by induced current in coil.
Mass	kilogram	Strain gauge in known gravity.
Pressure	bar $- 10^5\,N\,m^{-2}$ pascal $= N\,m^{-2}$	Strain gauge, Bourdon tube, diaphragm capacitive change, piezoelectric crystal.
Power	watt $=$ joule s^{-1}	Energy measured and time.
Sound Intensity	dB, dBA	Microphone, ultrasonic microphone.
Strain	% or metre	Strain gauge (change in resistor geometry).
Temperature	kelvin	Thermocouple, Thermistor, Semiconductor (V_{be}).
Time	second	Counter.
Torque	newton metres	Strain and displacement.
Velocity Fluid velocity	$m\,s^{-1}$ $m\,s^{-1}$	Displacement and time. Venturi pressure, wand/strain gauge; see flow.
Vibration	s^{-1}	Counter (and displacement).
Viscosity	$kg\,m^{-1}\,s^{-1}$	Differential flow, drag plate, ultrasonic.

The third measurement technique is more complex. In some cases a transducer may not be available which covers a sufficiently wide range of values and only allows a single 'null' point to be measured. Consider, for example, measuring the liquid level with a single light source and sensor. A tube could be connected to the tank so that a 'sample' of the liquid can be admitted filling it up to the same level. A valve could then be opened to release the sample. The time taken for the sample to run away until the sensor indicated that the tube was empty would indicate the depth. This is a **System** or **Servo** transducer measurement as it requires a perturbation of the system to detect a 'null'. This perturbation is likely to affect the value being measured more than the intrusion of direct or indirect transducers.

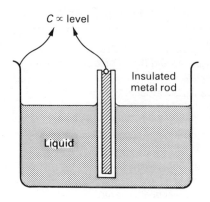
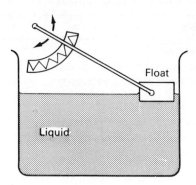

Fig. 2.1 Direct and indirect transduction of a variable

The second division is into **Digital** or **Analog** form. Transducers which give continuous analog values, like both those in Fig. 2.1, will require a conversion stage to follow to give discrete digital values. Transducers which can provide digital encoded values at their output are obviously preferable as they can be interfaced more directly to the computer. This split is currently about 50/50, though under 30% of transducers were digital in 1979.

The final division is into **Absolute** or **Incremental** encoding. A transducer which can be read in isolation and yet gives the variables' value is an absolute encoder. For example stones are placed every mile from the centre of London radiating out along the major highways. Each stone is marked (in Roman numerals) with its distance from the centre and arriving at a stone with no other information tells you where you are. A direction arrow saves you walking a mile the wrong way! As an example, the optical disc encoder, discussed in Section 2.4 and shown in Fig. 2.5, is a direct digital absolute transducer of rotational position. When the motorways were built, however, it was still decided to put a post every mile but to save money they were not labelled. Now starting from the beginning of the road and counting the markers you still have the same information and accuracy but joining in at any isolated point you do not. Also there is no directional information to be had even by walking to the next post, unless they are painted with patterns in groups of at least three. This is an incremental encoding scheme examples of which are shown later in Figs 2.6 to 2.8 and are also described in Section 2.4. The saving that was made by not carving the stones is made in reduced wiring and inputs to the computer. Only a single input bit is needed, or only two bits if direction is required. This rises to three if a starting point is included as well. This compares with the full number of bits for the required range of an absolute encoder.

After choosing the variable to be 'transduced' and the measurement technique, the accuracy and any limitations to it need to be considered.

2.2 Transducer Accuracy

The total accuracy of a transducer, expressed as a percentage of the full range, is the measure of the ability of the transducer to be within acceptable tolerance whilst in operation. The final accuracy is a sum of errors caused in four main different ways, some of which may be calibrated for. These may be either absolute errors or incremental errors and manufacturers usually manage to quote whichever appears better! They will be explained using a clinical thermometer as an example.

First there are three definitions which concern the ability of the transducer to measure the desired value of the variable. These are

THRESHOLD, RESOLUTION and FULL RANGE.

The **threshold** is the minimum measurable value, i.e. the lowest point at which a reading can be obtained. This is 94°F (308 K) for a clinical thermometer.

The **resolution** is the minimum measurable value difference, which can be thought of as the 'quantization step' of the transducer. This is the first determinant of the accuracy and for the clinical thermometer is 0.1°F (0.05 K).

The **full range** is the maximum measurable value, i.e. the limit above which readings can not be obtained. This is 108°F (316 K) for the thermometer.

Secondly there are two factors which are the effect of the relationship between the input variable and output electrical signal being non-uniform:

MONOTONICITY and LINEARITY.

A transducer is **monotonic** if for all increases in values of the input variable the output signal either remains the same or increases, but does not decrease. The same applies in the reverse direction. The importance of this is that if the relationship did have a 'dip' in the curve (i.e. non-monotonic) and the algorithm using the values was looking for the direction of changes on which to take some action, it would fail. For example if we always heat up if the temperature is falling and cool down if it is rising, non-monotonic transducer behaviour would cause us to switch from cooling to heating even though the temperature was rising. The transducer is still of use but a more complex algorithm would be needed.

The **linearity** is measured as the maximum deviation of the actual output signal from a straight line joining the end points of the range, and should be expressed in the same way as other accuracy measures (% error of full range). Many transducers, for example thermo-couples, are non-linear in operation but this does not prevent their use as it can be corrected for. The side effects of doing this are discussed under signal conditioning and signal conversion.

The example clinical thermometer is both monotonic, as mercury only expands if heated, and very linear as is shown by the uniform gradations on the glass tube.

The third group of errors are those which relate to the longer-term ability of a transducer to give consistent results.

HYSTERESIS, STABILITY and REPEATABILITY.

If readings are taken with the variable increasing and decreasing then there may be a difference. Often referred to as 'backlash', though more correctly as **hysteresis**, this is the difference between the readings for the same actual input. Our clinical thermometer is a poor example for explaining this as it is deliberately designed to take a maximum reading in the increasing direction and cannot be used in the decreasing direction at all. A rotational position sensor of any kind, connected to a shaft by a gear train, will exhibit hysteresis due to the air gaps between the teeth of the gears. Also many transducers using magnetic principles will exhibit hysteresis due to remanent magnetism being left while the variable decreases. Normally, any known 'hysterical' behaviour in a transducer can be corrected by the computer.

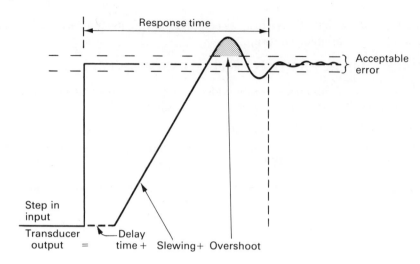

Fig. 2.2 Transducer response time

The **repeatability** of a transducer is an error between readings of the same value of input taken some small time apart. The **stability** is the repeatability of readings over a long period of time and is often quoted at the threshold and full range points. Drift in output for a given input can be caused by many physical processes, but variations in temperature (for all sensors except temperature sensors) and fluctuations in power supply for passive transducers are common culprits. Careful design is desirable to minimize these effects as calibration once in operation is the only alternative to maintain accuracy.

The final set of errors are all linked to time. Obviously, no transducer can respond instantly to a step in its input and its response is usually formed as shown in Fig. 2.2. An initial delay is followed by a period of **slewing** defined by a maximum slew rate quoted in volts per second and caused by the finite principles of operation of each transducer. This is reasonably predictable (and measurable) giving a direct relationship between the maximum change in variable output at any given frequency.

$$\text{Slew rate} = \text{maximum amplitude} . 2 . \pi . \text{frequency}$$

Most physical transducers have some mass and hence inertia which can give an initial delay but more importantly can cause an overshoot beyond the correct output value when slewing fast. The alternative is an asymptotic approach to the value and this is usually slower to settle within the resolution of the transducer, at which point the 'Response Time' has passed and the transducer gives the correct output.

2.3 Basic Transducers

Perhaps the simplest of transducers, but very common, are switches, often as micro switches for position or limit indications. Any mechanical contact such as a switch has a problem due to the inertia of the moving part when its position is changed. When the contact closes it 'bounces' off the fixed part, re-opening then closing again etc. This can continue for a few milliseconds and with a high-speed counter a single real closure can be counted as say 100 'bounces'. To get over this problem, there must be a switch 'debouncing' circuit or software.

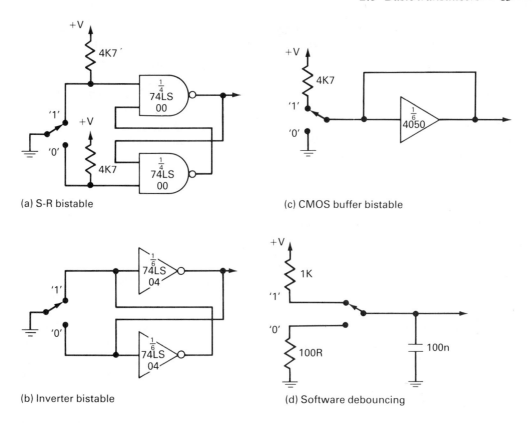

(a) S-R bistable

(c) CMOS buffer bistable

(b) Inverter bistable

(d) Software debouncing

Fig. 2.3 Switch debouncing

The traditional debounce hardware is a set/reset bistable (S-R) as shown in Fig. 2.3a. This operates by remaining set (or reset) until the switch contact reaches the opposite pole when the bistable output changes and the zero input is no longer necessary to hold it in its new state. Of course this event (like all debouncers) fails if the bounce is shorter than the set-time (two gate delays) of the bistable, but this does not happen with switches as the bounce is too slow and modern gate delays too short. This is not a cost-efficient circuit as for LSTTL it requires two resistors and two gates with a minimum of three pins on an integrated circuit. LSTTL outputs cannot be shorted without the possibility of damage, but in the circuit of Fig. 2.3b, an 'inverter bistable', this shorting only occurs for the switching time of the bistable which is quite safe. This needs only two inverters and a minimum of two pins on an IC. Any bistable element can perform the debounce function and so an LSTTL buffer with a feedback resistor will do, but it is not really cheaper or more convenient. However if MOS (Metal Oxide Silicon) circuits are used a real saving can be made as usually their outputs can be shorted without damage. Figure 2.3c shows the circuit which requires only the single buffer (amplifier) with direct feedback path and hence only a single pin! It may not be obvious how a single pin circuit can operate but it is quite simple. All that is necessary is that it can clamp the line to one state or the other in the absence of any input (switch changing or bouncing) and that the circuit can have its input pulled up or down (by the switch) with the same action, over-riding the current supplied by the output (which holds the previous state) and not damaging the output elements.

The microprocessor's software can be used to debounce switches instead of the external circuits described before if the extra instructions and time are acceptable. A switch connected to an input, such as in Chapter 12, can be connected to ground for the zero, but will need a resistor to limit the current from the supply rail, to indicate 'one', in case of failure. The additional components in Fig. 2.3d are only needed if it is desired to limit transitions such that they do not reverse in direction during switching. The software looks at the switch input(s) until a change is detected. These inputs are then ignored for a few milliseconds to allow for bounce before the final new state is taken as valid.

Many transducers are based on variations of resistance or capacitance due to an external input which gives the electrical signal directly. A resistive element, Fig. 2.4a, can exhibit

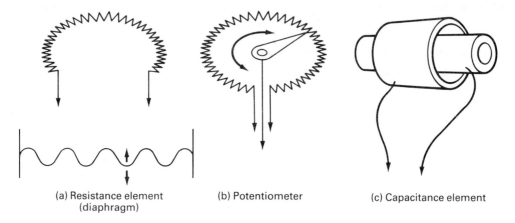

(a) Resistance element   (b) Potentiometer   (c) Capacitance element
    (diaphragm)

Fig. 2.4 Resistance and capacitance transducers

output changes due to alterations in its dimensions, the principle of the strain gauge described in Section 2.6, or alterations due to temperature, described in Section 2.5. The addition of a moving wiper, Fig. 2.4b, gives a potentiometer which can give position information as described in Section 2.4. Similarly a capacitive element as in Fig. 2.4c could have one fixed plate and one moving plate giving positional information; alternatively the composition of the dielectric, the material between the plates, could alter thus giving position (level) sensing, composition or flow information. The capacitor would either be connected in the feedback path of an operational amplifier to give a voltage proportional to 'C' or better be used to control the frequency of an oscillator.

As there are far too many principles of operation to include them all, four very common groups of variables have been chosen for a more detailed view. Position (displacement), temperature, pressure and flow between them cover a surprisingly large number of applications!

2.4 Position Sensing

A single position can be sensed by a micro-switch as previously described, but of wider application are sensors which can give a range of linear or angular positions to a very high accuracy. Both absolute and incremental, digital and analog position transducers exist and all are commonly used.

Potentiometers, both linear and rotary, are examples of analog, absolute devices, Fig. 2.4b. They are constructed by winding platinum alloy or other resistance wire round a former or by depositing a layer of plastic conductive material on a substrate. A wiper then runs along a track line on the surface of either to give a moving centre tap. If a voltage is applied across the two ends of the track then a voltage proportional to the position of the centre tap can be taken from it and one end. A hybrid construction is available which has the wire wound form coated with a conductive plastic. Wire wound pots are more stable and linear but their resolution is limited by the number of turns of wire fitted side by side. This could be as much as 20 turns per degree of rotation but the wire would be very thin! The carbon based plastic film on the other hand has its resolution limited only by particle size, but the films are not so linear and their resistance varies with temperature. The hybrid approach attempts to get the very best of both worlds by using a wire wound base and a conductive film to interpolate (and fill) between neighbouring wires.

One problem common to potentiometers and all devices with wipers is that at higher velocities the wipers bounce off any surface irregularity. A limit somewhere between one and ten thousand degrees of rotation per second is typical for very high-quality potentiometers. All potentiometers are passive. They may be energized by direct or alternating current, in which case they may form part of an oscillator circuit. This would give a frequency output rather than a direct voltage output, in a somewhat similar fashion to the circuit shown in Fig. 6.7. Potentiometers are cheap, simple, fairly reliable and continue to be used despite the availability of vastly superior optical transducers described below.

For completeness an analog incremental transducer is included here. A **synchro-resolver** works using a set of armature coils (rotor) and fixed coils to generate sine and cosine electrical signals which are then converted to give the relative position and direction of rotation. The basic principle is identical to a bicycle dynamo. Synchro-resolvers were commonly used in analog control systems as a great improvement over potentiometers but they are complex to make and can never approach the accuracies available from optical grating encoders. They also usually require slip rings to carry the energizing current to the rotor.

A direct digital reading could be achieved (and was) by having slip wipers running on a metal disc patterned with conductive and insulated parts. The encoding could take whatever form was preferred and digital output would result. It would be an improvement if positions could be detected without needing any contact (i.e. no wipers) between the item being sensed and the sensor. This would give a non-invasive transducer. Replacing the metal disc with a transparent one and having opaque patches to give the digital encoding will provide this. The addition of light sources and sensors completes the transducer as shown in Fig. 2.5.

To digress for a moment, the analog potentiometer could be replaced by a similar optical method using a photoresistive material instead of the conductive plastic. This, with a light source and a moving slot to replace the moving wiper, would give a 'photopotentiometer' capable of higher speeds with lower noise operation.

To return to the optical disc encoder, the light sources are light-emitting diodes (LEDs) and the sensors usually photodiodes, the pairs being matched to the same wavelength, often infra red (of about 900 nanometres) to increase the coupling efficiency. The resolution would appear at first sight to be fixed either by how close markings can be made on the disc, the size of the source and sensor or the amount of light passed through a slot between two adjacent markings. A review of the possible encoding schemes will elucidate which is the real limit.

If the disc were to be encoded in the most convenient form for subsequent processing,

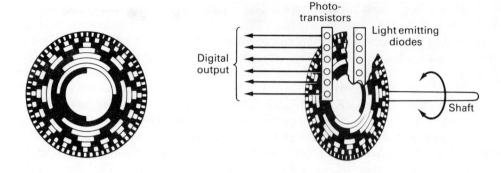

Fig. 2.5 Optical disc encoder

Binary encoding

Gray encoding

Incremental encoding

Fig. 2.6 Encoding patterns

pure **binary** as in Fig. 2.6a, then an absolute digital encoder would result. It would, however, be non-monotonic and give rise to false codes on transitions. For instance between binary **3** and **4**, either codes **0** or **7** could result.

Discs for absolute measure are almost always coded in one of the Gray codes which are monotonic and often referred to as reflected codes, Fig. 2.6b. This coding only permits one bit to change between any adjacent codes and gives a system where removing the top bit leaves a mirror reflection of the two halves, hence the name. The disc in Fig. 2.5 is encoded in this way.

The limit to resolution of this type of encoding is not set by how finely we can produce the grating lines but by the size of the source and sensor and the collimation of light to pass through only 1 coded slot. Accuracies of more than 8–10 bits are not easily achieved even though gratings up to 1 m in length and to 100 lines/mm can be made giving a possibility of over 20 bits. To take advantage of the excellence with which gratings can be produced a system is needed which will allow the beam between a source and sensor to overlap many (hundreds) of slots yet still give resolution to a single slot. This is not practical with absolute

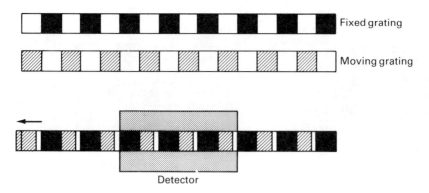

Fig. 2.7 Moiré fringe encoding

encoding but only with an incremental transducer. A simple **incremental** encoder, with the same resolution as the absolute, needs only a single source and sensor, as shown in Fig. 2.6c. A counter must be used to keep track of the absolute position. The addition of a second identical grating means that the entire area will alternate between light and no light as the moving grating passes the fixed one. Figure 2.7 shows how this allows the source and sensor to be as long as we like, within reason, compared to the size of an individual grating mark. This gives the simplest of the Moiré fringe techniques. It does rely, like the motorway mile posts, on a knowledge of the start point and does not have any directional sense.

The addition of a second sensor slot to an ordinary incremental encoder ninety degrees out of phase with the first will provide direction sense. As the two signals are out of phase they will go on and off at different times. The order of the transitions indicates the direction of movement. An improvement on this would be to have four slots and sensors with a phase angle of ninety degrees between each. Then taking the first ANDed with the inverse of the third, similarly for the second and fourth, would remove small alignment errors. The overall problem with this particular approach is still that large sensors require precise alignment.

As the grating can be made with such precision a much better approach is available. If either of the gratings has a second row of markings set ninety degrees shifted from the original row, the other grating being uniform, then two large sources and sensors with imprecise alignment give the full accuracy and directional sense. This is shown in Fig. 2.8 and is the simplest full Moiré fringe encoder. It still only uses a single 'axis' for achieving its resolution. The addition of a second 'axis' to interpolate between an ON and OFF of the grating system can give us very much greater accuracy.

If the two gratings are arranged so that the single rows of lines are not parallel but set at an angle to each other, then as the moving grating passes the fixed grating the light/dark fringes move in the transverse direction. For a disc encoder like this a fringe moves from the centre to the periphery for each movement of one grating division in one direction. It goes inwards if the direction of movement is reversed. Two sensors give us direction and larger numbers further increase the resolution by subdividing the limit of accuracy of the grating lines. This full Moiré fringe encoding technique can be arranged to give up to 1 metre direct linear measurement and sub-micron resolution, radially 0.2 arc seconds, i.e. about 22 bits. This is by far the most accurate device described anywhere in this book. The optical coupling can work up in the Mega-Hertz frequency range.

Proximity and **Level** sensing both fit into the position group. Though important as a group

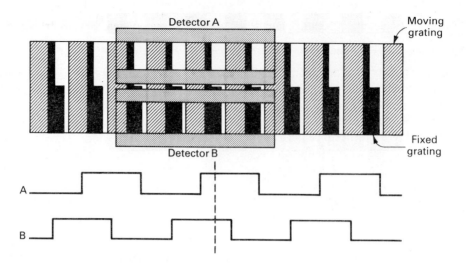

Fig. 2.8 Moiré fringe with directional sense

of sensors there is a wide choice of simple techniques, well described in the literature. Proximity sensors are based on contact by microswitches, ultrasonic beam interruption, capacitive, inductive or magnetic changes due to movement of the sensed item, infra red or visible light beam interruption or reflection. These techniques cope with detecting the proximity of almost anything. Level can be detected using a float (direct or magnetic linked) and any position sensor, by capacitive or inductive techniques such as Fig. 6.7, by liquid pressure (see pressure sensors), or by the various beam interruptions, particularly for opaque liquids. Further displacement measurement transducers are discussed with pressure sensors.

2.5 Temperature Sensing

Any thermal effect can be used as the basis for a temperature sensor, although only some are practical. The first and perhaps the simplest temperature transducer to explain, though now superceded, is the bimetallic strip. This analog indirect transducer consists of two parallel strips of metal with different coefficients of thermal expansion. One end of them is fixed and the other is free to move and connected to a position sensor. Alterations of temperature then cause the bimetal strip to bend and the amount of bend is proportional to the temperature. A range of 250 to 750 kelvin is achievable but with limited accuracy.

The second type, however, is in common use for higher temperatures, those of thousands of kelvin. The **Optical pyrometer** works by comparing the intensity of light emitted by hot materials. Iron, for example, varies in colour from dull red to bright white but using a single colour filter (usually red) to pass only a narrow spectral band allows just the intensity to be compared with an adjustable filament which has been previously calibrated. No direct connection to the hot material is required, only the optical coupling with sensors for the source and reference. This system-type transducer can give resolutions to 0.1 kelvin. Calibration to the 'gold point' (which is 1336 kelvin, equilibrium between liquid and solid) allows the relative brightness of any other temperature to be found from Planck's law. A less accurate method is to measure the intensity directly with a single optical sensor.

For the more usual temperature range there are three common temperature transducers:

Thermocouples, Resistance Temperature Devices and **Semiconductor Devices**.

The most widely used are the thermocouples. They give a range of 250 to 2000 kelvin and are rugged and reliable. They do, however, suffer from two serious drawbacks: the need for a reference junction and poor linearity. They provide a passive e.m.f. generator by the Thompson/Seebeck effect at the junctions of two dissimilar metals, the voltage being proportional to the difference in temperature of the junctions. The voltage relationship is not linear over the entire range and is also very small, only 0.01–0.04 mV/K for various metal pairs. One junction must be held at a fixed reference temperature as otherwise there would be two varying voltages, and a complex relationship between them. Also, if say Rhenium and Tungsten are used then at some point each will have to join the copper of a printed circuit board or other circuit components. Both of these junctions will exhibit the effect and thus these points must be held at a constant temperature (reference or 'cold' junction) to fix any offset error. An ice bath used to be a common reference technique though an isothermal block with electronic heating control and feedback is possible now. A common alternative is a block to retain the two reference junctions (X-Cu and Y-Cu) at the same temperature and to connect one in series with a self-compensating bridge including a resistance-temperature device to allow for alterations in reference junction temperature. For lower accuracy industrial uses an isothermal block at ambient temperature without a bridge compensating circuit provides adequate performance. Correction of linearity errors is discussed in Chapter 4 and Table 2.2 lists common thermocouple types.

Resistance Temperature Devices (RTDs) operate by the changes in resistance due to temperature of two broad groups: metallic conductors, and semiconducting metal oxides (thermistors). The metallic conductors are often arranged wire wound on a former and offer high stability and reasonable linearity over a wide temperature range. They do have a long thermal time constant and hence a low frequency range. Thermistors on the other hand are faster, but operate over a reduced temperature range. Fabricated from sintered metal alloy mixtures they have a large (negative) temperature coefficient and are more sensitive yet less accurate. They normally have a non-linear response requiring correction as described in Chapter 4. A comparison of the properties of three RTDs is shown in Table 2.3.

For use with a modern microprocessor system, and particularly as the 'sharp end' of an integrated transducer system the temperature sensitivity of the base emitter voltage of a

Table 2.2 Thermocouples

+ Metal Pair	−	Range K (approx.)	Features	Type
W (Tungsten)	Re (Rhenium)	270–2250	High temp, medium output	C
Fe (Iron)	CuNi 60/40 Constantan	25–1250	Economical, high output	J
NiCr 80/20 Chromel*	NiAl 97/3 Alumel*	3–1650	High output, very linear	K
PtRh 90/10 (Rhodium)	Pt (Platinum)	220–2000	Corrosion resistant, high temp, low output	S
Cu (Copper)	Constantan	3–650	High output, low cost, also humidity sensor	T

*Hoskins alloys Chromel and Alumel contain traces of other elements notably Si, Mn or Fe.

Table 2.3 Resistance temperature devices

Material	Range K (approx.)	Coefficient (accuracy)	Features
Pt wire	10–1100	0.04%/K (0.01–0.1%)	Fairly linear, slow response, corrosion resistant, stable
Ni wire	50–1000	0.5%/K	Non-linear at low temps, slow, stable
Thermistor	175–600	1–5%/K	Fast, non-linear, cheap, high output and sensitivity

transistor (similarly for diodes) is an ideal effect. A combined temperature and pressure sensor on the same substrate is now quite practical too. The temperature coefficient of silicon is only useful and linear in a range of 220–500 K, above which other materials such as doped gallium arsenide are used. The use of two matched transistors, integrated on the same chip, gives as simple, accurate, reference-free transducer with long-term stability. Unlike the single silicon device which requires an accurate constant current source, only a stable ratio is required. For a single transistor:

$$V_{be} = \left(\frac{KT}{q}\right) \log_e \left(\frac{I_c}{I_s}\right)$$

if I_c (the collector current) is very much greater than I_s (the reverse saturation current). For a well-matched pair at the same temperature the difference in base emitter voltages is given by

$$V_{be} = \left(\frac{KT}{q}\right) \log_e \left(\frac{I_{c_2}}{I_{c_1}}\right)$$

where K is Boltzmann's constant, q is the electron charge and I_{c_2}/I_{c_1} is the ratio of the collector currents through the two transistors. Simply fixing the ratio and using an operational amplifier (Section 4.2.1) to detect the difference in V_{be}s gives a solid state sensor with fixed calibration, and accuracy to a part of a degree. With both single and pair sensors it is common to integrate the necessary electronics so that either a temperature proportional voltage generator or a constant, temperature-variant current source appears at the output. These extra circuits then have components trimmed (by laser) to give an exact relationship, for example one micro-amp per kelvin from a range of 220–425 K.

Much research is being pursued into silicon and other semiconductor transducers and is covered in more detail in contemporary papers.

All of the temperature transducers discussed so far are analog. An example of a digital device exists using the colour change of specially prepared crystals at fixed temperatures. Each mix changes at a given temperature so having a separate area for each step of resolution gives a digital thermometer. They are not very accurate and have poor long-term stability at present, but again research aimed at direct digital temperature sensing would be very time saving if successful.

2.6 Pressure Sensing

All pressure transducers are indirect in that the pressure can only be measured as a force per known area by the deflection or temperature change it may cause. Force and pressure transducers are, however, the basis for other measurements such as fluid flow or torque. As for any indirect transducer two elements must be studied: the primary conversion of pressure to displacement, and the secondary displacement to electrical signal, which are summarized below.

Primary	**Secondary**
(Pressure-Displacement)	
Piston	Capacitive
Diaphragm, Bellows	Inductive (LVDT)
Integral Diaphragm Structure	Other position sensors
Bourdon tube	Piezoelectric
Direct attachment (vessel strain)	Strain gauge (piezoresistive)

Running through the primary parts first, pistons are likely to be unreliable as they have sliding parts and problems with sealing, particularly at high pressures. Bellows are really extended diaphragms to give greater sensitivity and both can only operate at relatively low pressures. They simply deflect when pressure is applied to one face. Bourdon tubes are spirals or parts of them (C-shaped), twisted or helical (Fig. 2.9) arrangements of (usually) metal tubing sealed at one end. The other end is open to the fluid pressure source and any increase in pressure will cause a straightening out of the tube and a corresponding movement at the sealed tip. As any material can be used for the tube very high pressures or high sensitivity can be achieved. All these 'mechanical' primary transducers are slow in response.

If the vessel containing the fluid will exhibit noticeable strain when the internal pressure is increased then a strain gauge can be used directly mounted onto the vessel. Strain gauges are ubiquitous in that by measuring strain they can be used for pressure, force, torque and other measurements quite easily. They are very common and are discussed in detail later.

The final method of connection to the pressure source is to use a piezoelectric (PE) crystal which is an integral diaphragm structure. If a force is applied across two faces of the crystal an electrostatic charge is produced across another pair of faces. The crystals are either natural such as quartz and Rochelle salt or ceramics such as Barium Titanate and Lead Zirconate Titanate (PZT). The PE constant is defined as the ratio of charge produced to force applied (both per unit area) for each pair of axes through the crystal, and the materials used are those exhibiting a strong piezoelectrical sensitivity for one pair.

The output voltage varies depending on whether the crystal is compressed or sheared. Very high voltages, suitable for igniting domestic gas, are generated by compression but the

Fig. 2.9 Bourdon tube indirect pressure transducer

voltages generated by shear are more suited to transducers. The effect is of course like a capacitor which, when the pressure on it is altered, changes its charge, the problem being to measure the voltage across the 'capacitor' without discharging it. Thus steady state readings are difficult whilst alternating pressure waves are easy.

The piezoelectric effect also operates in the reverse direction since if a voltage is applied the crystal will vibrate. This is the basis of ultrasonic pressure wave transmitters and receivers and allows easy matching for frequency characteristics.

Now considering the secondary parts of pressure sensors, displacement transducers, any of those discussed in Sections 2.3 and 2.4 could be used. In addition the use of inductive changes usually in the form of movement of a ferrite slug in a **linear variable differential transformer** (LVDT) has been common in the past.

When pressure sensing is mentioned the transducer which most naturally springs to mind is the **strain gauge**. This can be used in a wide variety of ways on pressure vessels, diaphragms or compression tubes, and singly or in bridge (4-sided) configurations. The principle of operation is that the resistance of an electrical conductor changes if the length of the conductor or either of the other dimensions is changed, for example as a result of deformation caused by an applied stress. The stress causes a strain of l/L (alteration 'l' of original 'L') and this is related to the ratio of change of resistance to original resistance r/R by:

$$\frac{r}{R} = k \cdot \frac{l}{L} \qquad \text{where '}k\text{' is called the strain gauge factor.}$$

The strained material is either attached only at its ends (unbonded) or is bonded onto a vessel or diaphragm. The former are less rugged but can operate at higher temperatures as there is no glue to melt! The material itself is either wire, foil-etched into various patterns or semiconductor, usually doped silicon. The base material of the strain gauge will normally be temperature matched to the material to which it is to be attached, to ensure that temperature changes causing expansion or contraction of the sample do not give rise to strain in the gauge. Note, however, that the bridge connections discussed below also remove temperature sensitivity. Strains measured in this way give reliable measurement for up to 5% deformation typically, but of course the elastic limit must never be passed or a permanent error will result.

A typical, traditional foil gauge might be fabricated from CuNi alloy, have 18 mm length, nominal resistance of 120 ohms with a gauge factor of 2 and work over a temperature range of 240–370 K but with a temperature coefficient of about 0.05% per degree.

Strain gauges are often connected in bridge arrangements so that the resistance can be cancelled out leaving a direct relationship between voltages and strain.

$$\frac{V_{out}}{V_{ref}} = k \cdot \frac{l}{L}$$

When measuring compressive loads it is quite common practice to increase the sensitivity and accuracy of measurement by using similar gauges for the four arms of the bridge and mounting them as shown in Fig. 2.10 to make use of both the contraction in the direction of the applied stress and the expansion perpendicular to it (Poisson's Ratio). A similar arrangement can be made for tension.

The use of silicon in modern strain gauges allows a complete gauge or diaphragm to be made with four etched or ion implanted resistors in a bridge to give accuracies better than

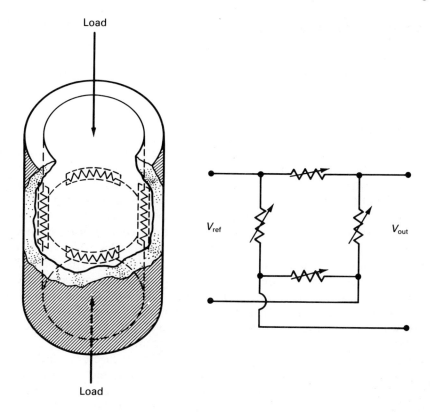

Fig. 2.10 Strain gauge, compressive bridge connection

Table 2.4 Pressure sensors

Type	Range (approx.)	Accuracy	Features
Diaphragm	0.1 mb–100 bar	0.1%	flush mounting, simple capacitive circuit
Bourdon	25 mb–10 kbar	0.5%	sensitive or rugged construction, wide range
Strain Gauge	25 mb–5 kbar	0.2%	higher cost, fairly wide temperature range
Gauge (Semiconductor)	5 mb–1 kbar	0.1%	high frequency (50 kHz) large temperature dependence if single
Piezoelectric	5 mb–5 kbar	1%	wide temperature range, high frequency, active

0.1% overall. As silicon 'resistors' can be made p- or n-type (respectively boron or phosphorus doped) and these have reverse sensitivities to strain, the bridge can have all four arms in tension and cancel out the higher temperature sensitivity that goes hand in hand with high gauge factors (more than 100). Again much research is currently being pursued into silicon transducers for pressure sensing and is well covered in the literature.

Table 2.4 shows a comparison of features of pressure sensors and for those more at home with imperial measures 100 bar is roughly 1500 psi and 10 kbar is roughly 65 tons per square inch.

There is one further point of importance for pressure sensors, that of their starting-point. Absolute reference transducers are evacuated, sealed and balanced to give a zero output at zero pressure absolute. This seems quite obvious but the confusingly named 'Gauge' reference transducers are vented to ambient pressure and balanced to give a zero output at atmospheric pressure.

2.7 Flow Sensing

The general term 'flow' covers three measurements: velocity or flow rate, volumetric flow and mass flow rate. As the volume flow is easily calculated from the flow rate, cross-sectional area and a knowledge of the type of flow, and mass flow requires only the addition of temperature and density to these, only flow rate measures are described here. There are various subtleties to the other measurements but these are well covered in the literature.

Traditionally flow rate was measured either by direct contact or differential pressure techniques. Both of these are invasive in that they obstruct the flow which is to be measured. Thermal transfer has been used but it is invasive in a different way as part of the fluid has to have its temperature altered. More recently both electromagnetic (inductive or capacitive) and ultrasonic transducers have started to take over as they are non-invasive and easier to connect and use. The latter hold great promise as integrated sensors. All flow meters need to be placed with care as if they are downstream of an obstruction or bend the velocity profile may be disturbed and cause incorrect readings to be taken.

Direct contact flowmeters are usually active, digital transducers in which a turbine or paddle wheel rotor contains a small permanent magnet. As the fluid passes the rotor, momentum transfer causes it to rotate and the magnet moves close by a fixed pickup coil giving a series of pulses. The pulse frequency is proportional to the flow rate. This type is commonly found in small yachts for measuring the flow of the ocean past the hull! More than one magnet can be included in the rotor and accuracies of better than 1% are common for flows of 5% to maximum. Problems occur at low flow rates due to friction and low pulse rates giving low response times.

Differential pressure sensors use ordinary pressure sensors to measure the difference in 'head' between two points in the stream. An orifice plate, Venturi or Pitot tube is usually the base for providing the two close points though for greater accuracy multipoint measurements should be made. This type of transducer is non-linear and also temperature dependent and an integral temperature sensor for compensation should be included. Gas flows at low pressure are very hard to measure by this technique.

Thermal transfer, as its name implies, includes a small heater upstream of a temperature sensor. Other temperature sensors are required to remove any ambient changes. The amount of temperature increase generated by a known input of energy is proportional, though highly non-linear, to the mass flow. This is a simple technique with no moving parts but has to be calibrated for the thermal characteristics of each fluid. The rate of removal of heat can be sensed as well as the rate of transfer to give better results.

The more recent developments in sensing provide no obstruction to the flow. Electromagnetic induction flow measurement is based on the e.m.f. which is generated by the flow in a magnetic field. An electromagnet is placed around the pipe carrying the fluid to produce this, and the field is usually arranged to alternate to remove any voltages generated by electrochemical or thermoelectric effects which also operate. This type of flow meter can only operate in a satisfactory fashion if the fluid is sufficiently conductive and this excludes many common liquids.

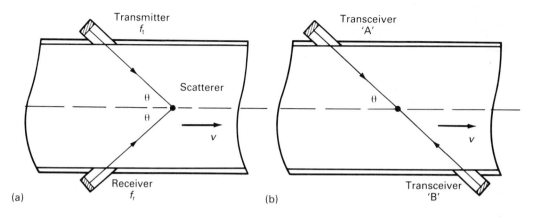

Fig. 2.11 (a) Doppler and (b) transit-time ultrasonic flowmeters

Future developments of non-invasive flowmeters are very likely to be dominated by ultrasonic techniques, for which two quite different principles are used, shown in Fig. 2.11. The flow rate can either be detected by the transit time of an ultrasonic wave being increased in the direction of flow (and decreased in the opposite direction) or by frequency change of a wave on being scattered when it hits a moving particle in the flow. This Doppler Shift method (a) thus requires scatterers in the fluid such as bubbles or impurities, but has a high degree of repeatability, though it is difficult to calibrate initially as the quantity and distribution of scatterers vary enormously. This technique is already used for yacht speed sensing. Transit time flow meters (b) operate on clean or impure fluids and using both direct (leading edge) and indirect (difference) time measurements can give highly accurate results. Only a small proportion of pulses transmitted in each direction will be needed to be received to determine the rate. The time differences are small (10s or 100s of nanoseconds) but modern electronics are highly suited to the task and this type of transducer offers the potential for great improvements when integrated as described in Chapter 1.

This chapter could not cover all of the enormous field of transducers, but there are some more details in the examples of Chapter 15. There has been no mention of the wide range of chemical analysis transducers, luminous intensity sensors, magnetic sensors, etc, etc. It has, I hope, whetted the appetite and given a direction into the technical literature which exists in profusion.

3

Input Sampling

By their very nature all computer-based data logging and control systems are sampled data systems. Any given transducer's data is not continuously monitored as the computer performs calculations, takes data from other transducers or outputs control signals. Hence the following questions should be asked:

(a) What is the actual information to be input?
(b) How often must the input be sampled?
(c) What is the effect of the finite time needed to take a sample?
(d) What noise is superimposed on the information signal?
(e) What errors are introduced by points (b) to (d) above?
(f) Is any additional circuitry necessary to alter the input signal in any way as a result of these considerations?
(g) Does the computer, for example by its word length, affect our consideration of the information we sample?

A brief introduction to frequency spectra is needed as it is easier to visualize any errors which may be caused by seeing pictures of frequencies rather than pictures in time. A single sine wave has a well-known shape (such as in Fig. 3.4) if plotted by amplitude against time. Using a frequency axis against maximum amplitude gives the graph in Fig. 3.1a. Any real signal is more complex than this and, if it is a continuous wave, can be analysed as a sum of various individual sine wave frequencies as in Fig. 3.1b. This is a mathematical model which is only correct if the waves may be considered to extend infinitely in time and so does not hold for isolated pulses. It will suffice for all normal analog transducer inputs.

Reverting to taking samples in time it is easily seen that the signal shown in Fig. 3.2 is how we sample. There is a relatively long period (T—the sample time) in between actually taking the samples for a relatively short period (τ the aperture time).

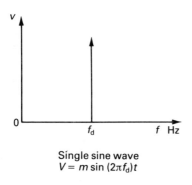

Single sine wave
$V = m \sin (2\pi f_d)t$

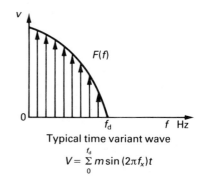

Typical time variant wave
$$V = \sum_{0}^{f_d} m \sin (2\pi f_x)t$$

Fig. 3.1 Simple frequency spectra

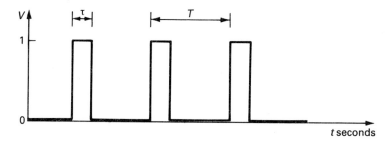

Fig. 3.2 Normalized sampling wave

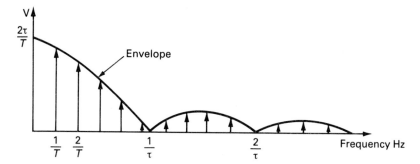

Fig. 3.3 Frequency spectrum of sampling wave

These are the two most important variables in any sampled system because, as will be shown, the sample time T provides a limit on the frequency of any signal which is to be sampled, and the aperture time τ provides a limit to the accuracy of measurement. The frequency spectrum of the normalized sampling wave is shown in Fig. 3.3. This is what is 'applied' to our input data by our sampling it, and so may cause error!

3.1 Sample Time (*T*)

A sine wave $a = A\sin(2\pi ft + \theta)$ has three variables: amplitude, frequency and phase. When it is sampled only the amplitude is taken, the frequency and phase being determined inherently by the times at which the amplitude samples are taken. A lengthy mathematical proof could be used to show the low limit of samples, below which this inherent data is lost, and derive the minimum sample rate or Nyquist criterion. I have chosen instead to show the effect of falling below this criterion and then demonstrate why this is so, to avoid unnecessary maths!

The Nyquist criterion is best stated as: 'if a signal has its highest frequency fHz then it is necessary to take more than $2f$ samples per second to enable the signal to be reconstructed', i.e. all data from the wave has been acquired. Considering some special cases shows this. If the wave is sampled exactly twice per cycle ($2f$) then the samples could all occur at the zero crossing points and so no data would be recorded! Looking at Fig. 3.4 the results of sampling at $8f$ and $4f$ show that only a single sine wave of frequency f or less could be drawn through the samples. Thus all the data has been captured. However, if the sample rate is dropped to $1.5f$ then it is easy to see that another wave at a lower frequency than the original can be

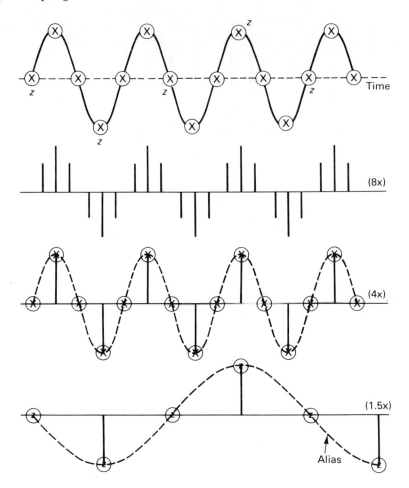

Fig. 3.4 Sine wave sampling at different rates

fitted to the samples. This is called an **Alias** and appears as a corruption to the original, as noise which cannot subsequently be removed if it overlaps any frequencies in the original. The alias signal frequency is the difference between the actual input signal frequency and some harmonic (multiple, $1 \ldots n$) of the sampling frequency. But to see why it happens one needs a closer look at the frequency spectrum of the original and sampled waves. The sampling signal, Fig. 3.3, may be analysed by traditional Fourier analysis techniques and if the signal is normalized to 1 volt then:

$$V(t) = \frac{\tau}{T} + \frac{2\tau}{T} \sum_{n=1}^{n=\infty} \frac{\sin(n\pi\tau/T)}{(n\pi\tau/T)} . \cos\left(\frac{2n\pi}{T}\right) t$$

If for simplicity the sampled signal is a single wave of frequency f with zero phase then:

$$V_s(t) = a . \sin(2\pi ft) . V(t)$$

$$V_s(t) = \frac{a.\tau}{T}\sin(2\pi ft) + \frac{2a.\tau}{T} \sum_{n=1}^{n=\infty} \frac{\sin(n\pi\tau/T)}{(n\pi\tau/T)} \left(\cos\left(\frac{2n\pi}{T}\right) t . \sin(2\pi ft) \right)$$

It can be seen that the signal after sampling has gained some extra components, of the form Sin A, Cos B, in a manner very similar to modulation for data transmission.

$$V_s(t) = \frac{a.\tau}{T}\sin(2\pi ft) + \frac{a.\tau}{T}\sum_{n=1}^{n=\infty}\frac{\sin(n\pi\tau/T)}{(n\pi\tau/T)}\left[\sin\left(2\pi\left(\frac{n}{T}+f\right)t\right) - \sin\left(2\pi\left(\frac{n}{T}-f\right)t\right)\right]$$

This is the original signal multiplied by a factor (τ/T), which is constant for all frequencies in the sampled signal plus some other terms. The factor (τ/T), having been chosen for the sampling signal, is corrected for by amplification inherent in the sampling electronics and the extra components. Now any real signal is of the complex form in Fig. 3.1b. The effect of this is that each of the $\sin(xf)$ terms becomes a series of terms from zero to f which simply converts a single line to a 'whole bunch' of lines up to f in the frequency plane. The first term in the equation above (now a series of terms) is still the original signal with the constant factor. This is shown in Fig. 3.5, as is the first term $(n = 1)$ of the rest of the equation. The other terms $n = 2$ to $n = \infty$ extend off the page to the right!

As long as $1/T > 2f$ then the terms are all separate in the frequency domain and the sampling rate is adequate. If, however, $1/T \leq 2f$ then frequency components from $1/2T$ to f are folded back on top of those from $(1/T-f)$ to $1/2T$. This is called frequency folding and is responsible for the signal errors, and alias production, as in Fig. 3.6.

The obvious solution is to sample at a frequency greater than the Nyquist criterion but

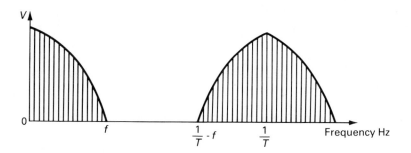

Fig. 3.5 Adequate sampling rate

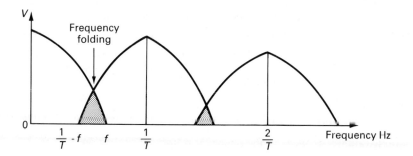

Fig. 3.6 Inadequate sampling rate

there is a problem: Noise! If, for example, we sampled at 20% above the Nyquist point and there was some superimposed noise at $3f/2$ on the sampled signal then due to frequency the noise will be folded about the sampling frequency and so distort the original signal at: $(2f + 20\%) - 3f/2 = 7f/10$. The higher the frequency of the noise (and HF noise is common as described in Chapter 10), the worse the problem!

Thus, the only satisfactory solution is a two-stage procedure. First use an adequate sampling frequency and second connect in a filter to remove higher frequency components from the sampled signal. This is called an **Alias Filter.**

An adequate sampling frequency is more than twice the maximum data signal frequency as the theoretical limit. Two and a half times f is the minimum practical rate used for continuous waves. Lengthy mathematics shows that for a single cycle, four samples is the minimum theoretical limit and five samples per cycle is the practical rate used. Higher rates are, of course, perfectly satisfactory.

There is a loophole for using lower rates in certain very special circumstances. If, after folding, the folded signals do not overlap the original signal, due to there being gaps in the spectrum of the original, then the folded parts can be removed and no error results. Another example is if a repetitive wave is available, e.g. your heartbeat when relaxing and reading this! Then a much slower sample rate can be used provided that the sampling frequency is not a sub-harmonic of the input wave, and the input can be slowly built up in the store by placing each sample in the correct place. This also averages out any random noise which may be present, and the sampling may be performed stochastically to assist this.

In the former case this applies to band-limited signals which require sampling. To fit the folded spectra into the gap(s) left by band limiting the original signal (limited from $f_1 \rightarrow f_2$) we simply require that:

$$\frac{2f_1}{K} > f_s > \frac{2f_2}{K+1} \qquad \text{where } f_s \text{ is the sampling frequency and } K \text{ is an integer.}$$

The lowest sampling frequency is found when K is a maximum which is when

$$K_{max} < \frac{f_1}{f_2 - f_1}$$

Now considering the alias filter, Fig. 3.7 shows what will happen if we do not filter out the noise. Two filter responses are shown and the ordinary CR filter is obviously inadequate. The response of a CR filter is 3 dB (deciBels) per octave, that is the input will have been cut to **half** its amplitude for each **doubling** of frequency. An active filter (i.e. one containing an operational amplifier) is essential if low sampling rates are to be used. Very complex active filters can be obtained with responses of 80 dB/octave and phase matching to 1/2% and so come close to the ideal 'vertical line' filter we would like.

For an extremely simplistic view of what happens think of the stagecoach in an old cowboy movie. Remember that it is filmed (sampled) at 24 frames per second. As the stage pulls out everything appears normal but as it speeds up the wheels appear to reverse and rotate backwards. The wheels are moving faster than half the sampling rate of 24/s, an alias is seen, and all information on the positions of the spokes, which are the highest frequency, is corrupted.

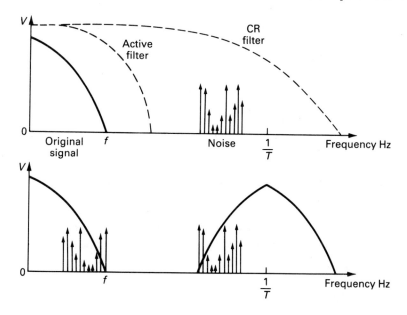

Fig. 3.7 Sampling in the presence of noise

3.2 Aperture Time (τ)

The frequency and phase of the input signal are inherently measured by knowing when the samples are taken. Now if samples are being taken often enough, what is the effect of the finite time to take a sample, and what is the effect of any uncertainty in the time at which the sample starts and stops?

The duration of taking a sample acts either as an averager over the aperture time or as an uncertainty in when the sample was taken depending on the electronics involved. If it is simply averaging then a loss in accuracy of amplitude, experienced as a more limited resolution, is all that happens. **Aperture Uncertainty** on the other hand affects the frequency and amplitude accuracies and so has a complex effect on the signal.

Referring back to the signal we have obtained by sampling the wave X:

$$V_s(t) = \left[\frac{\tau}{T} + 2\frac{\tau}{T}\sum_{n=1}^{n=\infty}\frac{\sin(n\pi\tau/T)}{(n\pi\tau/T)}.\cos\left(\frac{2n\pi}{T}\right)t\right].X$$

So the original signal (X) has been multiplied by some (error) factor. The (τ/T) at the beginning is simply a fixed amplitude alteration which can be corrected. Each separate spectral line at n/T ($n = 1$ to $n = \infty$) is a wave $\cos(2n\pi t/T)$ and we can simplify the calculation by taking the envelope joining the tips of these spectral lines to be a continuous error wave. Substitute the variable 'f' for (n/T) and the envelope of the multiplying factor becomes

$$V_{env}(f) = \frac{2\tau}{T}\left|\frac{\sin(\pi\tau f)}{(\pi\tau f)}\right|$$

The amplitude of each folded spectrum, all frequencies of $F^*(f)$, is attenuated by this as in Fig. 3.8. It contains the gain factor $(2\tau/T)$ and a frequency-dependent error factor, so the error can be extracted as:

$$\text{Error} = \frac{\Delta V}{V} = 1 - \left| \frac{\sin(\pi\tau f)}{(\pi\tau f)} \right|$$

This links error, aperture time and frequency and a few simple checks show it to be sensible. It tends to unity (total error) if $f = 1/\tau$. The aperture time is such as to take in a complete cycle of the sampled wave (highest frequency) and so the reading is averaged to zero. It tends to zero in two cases; if τ goes to zero we have infinitely short sampling and would expect no error, if f goes to zero we are sampling direct current and any averaging still gives no error, but fixed data.

The relationship between error, expressed both as a percentage and a word length in bits, highest frequency of the sampled signal and aperture time is plotted in Fig. 3.9.

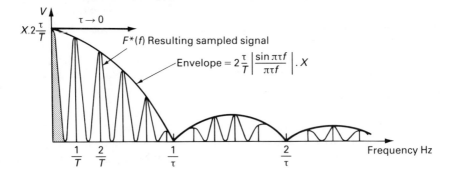

Fig. 3.8 Result of sampling original signal X

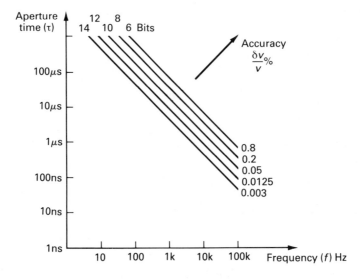

Fig. 3.9 Aperture time, frequency and accuracy (continuous)

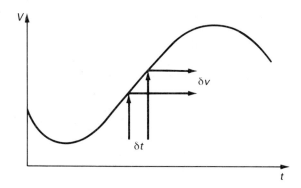

Fig. 3.10 Aperture time uncertainty causing error

Remember that this is the error for the whole signal and it must be a continuous signal for this analysis. Note also that there is no dependency on the sample time T.

There is another way to view the error caused by finite aperture time and because it gives worse results and so means more equipment is required it is almost always quoted by manufacturers. If one just takes a single sine wave of the highest frequency in the signal and finds its worst rate of change, then that can be used to find the error. Fig. 3.10 shows this relation between error and aperture.

$$\delta V = \frac{d}{dt}(V \sin(2\pi ft))\tau \qquad \text{so in the worst case}$$

$$\delta V = V.2\pi f.\tau \qquad \text{or}$$

$$\frac{\delta V}{V} = 2\pi f.\tau$$

This second relationship between error and aperture time is shown plotted for the same values as before in Fig. 3.11. The large difference in the two methods can be seen by comparing some specific (and easy to remember) values.

If the aperture time is one tenth of the maximum frequency then the full calculation gives an accuracy of about 5 bits.

If the aperture time is one hundredth of the maximum frequency then the full calculation gives an accuracy of about 10 bits and the quick (manufacturer's) calculation gives only 5 bits.

For a ratio of one thousandth full calculation gives about fifteen bits and quick worst case gives about 8 bits. It takes an aperture time of one ten thousandth of the maximum frequency to get an accuracy of about 11 bits with the quick worst case calculation.

Both methods seem (are) correct, so how does this wide discrepancy occur? If a sample is taken at the worst case point then the next sample **cannot** be taken at the equivalent point due to sample time restrictions. Thus if **each** individual sample must be guaranteed to have a stated accuracy then the worst case should be used. Under any other circumstances, such as averaging algorithms, use of digital filter algorithms etc. then the normal calculation is correct. Also if a complex signal has a highest frequency *f* the rate of change will usually be damped by the lower frequency components to less than the worst case.

Thus, the aperture time τ is the time required to sample a signal until its amplitude is stored (fixed) and is equivalent to an error in the measured amplitude of the signal. Aperture uncertainty is the time over which it is unknown when the sample was taken and is equivalent to a frequency-dependent error in the sample. The only cure for sampling high frequency signals is to use some external hardware to shorten these two times and this is discussed in Section 3.4.

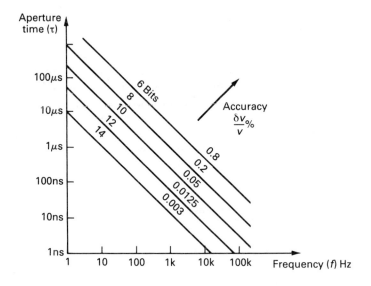

Fig. 3.11 Aperture time, frequency and accuracy (worst case)

Again for a simplistic view of τ-limitations consider taking a picture of a racing car with a fixed camera using an exposure time of 1/25th second. If the picture is taken at the correct time the car will be blurred (averaging due to long aperture). If the picture cannot be taken at a fixed time then the car may only partially appear or may be missed altogether (aperture uncertainty time). If the shutter speed is increased to 1/1000th second and opened at the right time then a 'perfect' picture will result as the aperture time is now right for the maximum frequency.

3.3 Word Length (*L*)

Any microprocessor converter we use will have a fixed word length and this, if it is short, may have limiting effects on sampling and subsequent processing of data. The separate problem of having a limited width data bus, say only 8 bits for a 16-bit internal machine, is discussed in the sections on signal conversion and interfacing to a computer.

There are three main effects of limited word length which will concern us:

(a) **Accuracy of each sample.**
(b) **Problems of small differences.**
(c) **Accuracy of constants.**

When each sample is taken, limiting its word length to eight bits limits accuracy to 0.2%. The effect is to introduce a more severe quantization error than may be present in the transducer and associated circuitry and this appears the same as the addition of this amount of random noise. This accuracy may prove quite adequate but, if not, a longer word length or double length working with consequently slower algorithms will be needed.

Remember that $\Delta V = V/2^n$ for an n bit word, i.e. the quantization threshold voltage for linear quantization.

All processing algorithms can be written in a number of different ways. Those which involve differences between numbers which are close together or some multiplications or divisions will result in a loss of accuracy, a loss of significant digits. This, again, is similar to the inclusion of some random noise but can be far more severe as the 'signal to noise' ratio may be so poor as to lose all 'signal' data. An example of this problem appears in Section 4.4.3. The solutions are either to choose a different algorithm, as in the digital filter example, to remove the problem or to use a longer word or double length arithmetic to retain the desired accuracy.

Constants required in any algorithm cannot necessarily be stored exactly, for example the coefficients in a quadratic equation, and the result of this can be critical. For example a 'pole' in the response of a closed loop control system may be shifted (by rounding error) so that the net result is a just unstable rather than a just stable system. Errors in constants could be cumulative and so the accuracy of data could suffer considerably with only minor inaccuracies in a number of constants. This problem is also discussed further with the algorithms of Section 4.4.3.

3.4 Sampling and Holding

A **short** aperture time (τ) during which the value of the signal must be determined is followed by a **long** sample time (T) between samples. If there was a mechanism to capture quickly and store an analog value then the whole of the sample time would be available for conversion. Conversely if the conversion of an analog signal to a digital form must all take place during the aperture time, the maximum frequency will be severely limited by available A–D converter rates.

Three cases when a 'Sample and Hold' circuit or the similar 'Track and Hold' would be useful are:

(a) The signal is high frequency and aperture time must be shorter than any possible converter could achieve.

(b) A slower converter is desirable to reduce cost or to simplify hardware.

(c) The signal is not continuously available, for example due to multiplexing. (See Section 4.3.)

To simplify the description of sample/holds, they will first be covered in concept and then as practical circuits. In the latter case, integrated circuits called 'operational amplifiers' are used and these are assumed at this stage but will be described in more detail in Section 4.2.1.

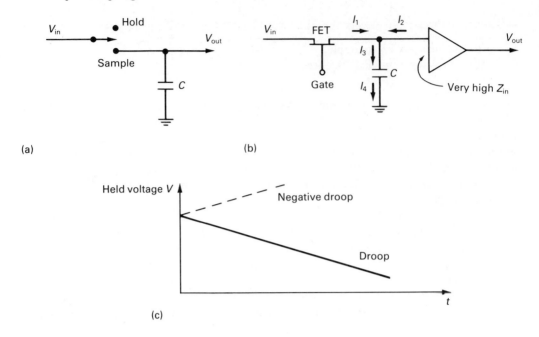

Fig. 3.12 Concept of sample/hold circuit

All electronic storage devices from simple bistables to the analog sample and hold stores being described here rely on the ability of a capacitor to hold a charge over a period of time. This holding of electrons allows a voltage to be 'captured' as $V = Q/C$ where C is the capacitance but as the stored charge (Q) will follow (sample or track) the applied voltage, a way of isolating the capacitor (hold) is needed. Figure 3.12a shows the simplest concept of the circuit where the input voltage can be connected to the capacitor or isolated from it by a switch. There are a number of problems shown up by this circuit which dominate practical S/H designs.

Firstly, the switch—any mechanical switch will be too slow as the whole purpose of the S/H is to cut aperture and aperture uncertainty times to nano- and pico-second values. An electronic switch is essential for sampling high-frequency signals. It needs to have zero ON resistance to ensure the voltage is sampled exactly and infinite OFF resistance to give complete isolation and obviously both of these will not be met. Secondly, the capacitor will not be perfect and will have some leakage of charge over time so the voltage cannot be held for long. Thirdly, the connection to the output, and hence to the analog-to-digital converter, will also cause alteration of the charge as a voltage can only be measured by a movement of charge and this will need to be as small a percentage of the original as can be arranged to keep the accuracy high.

So, moving to the somewhat more practical circuit of Fig. 3.12b, a Field Effect Transistor (FET) replaces the switch, having very high OFF resistance and reasonably low ON resistance. A logic signal to the gate of the FET switches the transistor on or off. Also to ensure minimal disturbance to the charge in the capacitor when it is checked an operational amplifier with a very high input impedance buffers the output.

There are now three currents to consider to decide if the charge in the capacitor will accurately hold the voltage sampled. I_4 is the inherent leakage of the capacitor due to the non-infinite impedance between the plates. Input to the capacitor is the sum of two currents due also to non-infinite impedances. I_1 is the leakage current through the OFF field effect transistor and I_2 is the leakage current of the buffer amplifier. If the capacitor leakage and its input are equal then the capacitor charge will remain static. In practice they are not and the capacitor voltage 'droops' with a droop rate usually specified in volts per second.

$$\text{Droop rate} = \frac{dV}{dt} = \frac{(I_3 - I_4)}{C}$$

This rate is defined as droop (i.e. a decrease) but of course I_3 could be greater than I_4 when the voltage on the capacitor will increase (negative droop!) as shown in Fig. 3.12c. It would seem the low droop rates can be achieved by incorporating a large capacitor but a consideration of the time response of the circuit will show that this is not possible.

As long as the switch is closed then the input voltage is connected (via the ON resistance of the FET) to the capacitor and the voltage on the capacitor should follow the input. It will not if sufficient charge cannot be transferred into the capacitor, or if the input cannot supply enough current.

When the switch is opened, which takes some time, the currently charged capacitor is isolated and the signal held. This time is the **Aperture** Time of the S/H and the uncertainty in its occurrence is the Aperture Uncertainty Time.

While the switch is open the signal is held but droops at the droop rate, and this is shown with the other times in Fig. 3.13a. The input continues to alter as the lower dashed line shows and at some point the logic signal commands the switch to close and to resample the input. The voltage on the capacitor has to alter to catch up to the input signal and can only slew at a maximum slew rate ($dV/dt = I/C$) and may overshoot before the capacitor voltage equals the input voltage. This is very similar to the response of transducers previously discussed. The total time for equalizing the voltages to within the desired accuracy is called the **Acquisition** time. To minimize it, the slew rate must be high and the capacitor correspondingly small, also the input current supply should be large, so that the impedance of the input should be low. This also allows the sampling phase to 'track' higher frequency signals. The circuit of Fig. 3.13b shows the inclusion of a unit-gain operational amplifier buffer to present a high impedance to the source and so affect it as little as possible, but a low impedance output to supply or sink current to the storage capacitor.

To recap on the requirements for a sample and hold circuit, it must have:

- High input impedance   —minimum effect on source.
- Short aperture time   —probably a fast MOSFET switch.
- Long hold time   —low droop rate.
- Short acquisition time   —high slew rate (hence fast settling time)
- High accuracy   —probably unit gain.
- Convenience of use   —good range and compatible logic signals.

There are special considerations for some sampling requirements. If only the maximum (or minimum) input voltage is required for a given period then adding a diode in series between the FET switch and the capacitor allows the tracking mode input to charge the capacitor only when

$$V_{in} > V_{cap} + V_{diode}$$

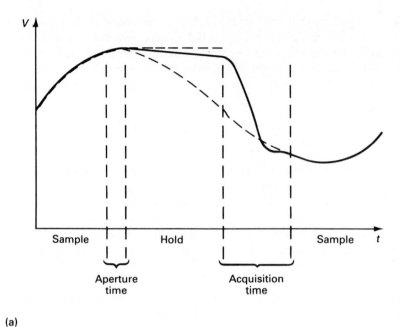

(a)

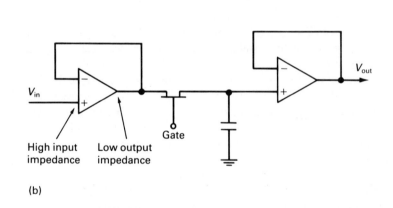

(b)

Fig. 3.13 Sample/hold circuit and timing

which with a simple reset circuit to discharge the storage capacitor after the sample has been used, gives the desired result. The sample and hold, as has been shown, averages the input voltage during the aperture time. It is sometimes desirable to obtain the sum of a number of samples, and adding an integrator into the S/H is one way of achieving this without requiring processing (Section 4.2.1) from a micro.

Some typical values for commercially available sample/hold circuits are given in Table 3.1; it should be noted that the most sophisticated units are expensive but have gain–bandwidth products extending well into the GigaHertz regions.

And so, answering the questions posed at the start of this chapter, the input must be sampled more than twice per cycle of the highest frequency for a continuous signal, more than four times per cycle for a single oscillation, unless the signal is such that there are gaps in its

Table 3.1 Typical parameters of commercial sample/holds

Type	Economy	General purpose	Fast	Ultra fast
Range	±10V	±10V	±10V	±5V
Aperture (ns)	200	100	50	10
Aperture uncertainty (ns)	10	5	2	0.05
Acquisition (μs)	20	4	1	0.2
Slew rate (V/μs)	0.5	5	20	500
Droop rate (V/s)	5	0.5	5	50
Linearity	0.1	0.01	0.01	0.05

frequency spectra into which the artefacts of sampling may be folded without distortion. An example of this latter case is a television signal. The effect of a finite time to take a sample is to limit the accuracy by averaging the input over this aperture time. The effect of any uncertainty in the time at which the sample is taken is a frequency-dependent error and limits the upper frequency which can be sampled. Methods of calculating frequency and accuracy for given apertures and uncertainties have been derived. If any higher frequency component or noise is superimposed on the information signal an **Alias** filter will be necessary to prevent frequency folding for any given sample time. If the aperture time or aperture uncertainty of the chosen analog-to-digital converter or conversion technique is too long for the input signal frequency and accuracy desired then a Sample and Hold circuit will be needed between the filter and converter. Adequate word length must be maintained throughout a system and this may require multiple precision words at some points. Algorithm and storage patterns should be carefully chosen to ensure no unnecessary loss of accuracy.

4

Signal Conditioning

Having seen where input signals are likely to come from and what the effects of sampling the signals will be it is still quite probable that some 'pre-processing' will be necessary to make the signals suitable for conversion or subsequent use in a microprocessor. For example it is unlikely that a steel mill, the home or a human brain would use the 0V–5V signals commonly used in the final computer interface. Any circuits (or programs) involved are grouped together under the heading Signal Conditioning and all the phases which may be included are shown in Fig. 4.1. The first two groups of circuits are concerned with isolating and amplifying or attenuating signals from transducers to make them match the voltage and current ranges of the standard electronics which one would wish to use throughout the rest of a system. These are initial **Ranging** circuits and they precede any necessary alias filter.

A choice is then available, whether the rest of the conditioning operations, mostly filtering and linearizing, are to be carried out with analog electronic circuits and signals or, after the conversion stage, by digital programs and storage. The former are called *pre*-conditioning and the latter *post*-conditioning. The intervening phase of analog-to-digital conversion is covered by Chapter 6 but the multiplexing of multiple signals to a single converter and the consequent need for sample/hold circuits is considered as part of the signal conditioning.

4.1 Isolation

The first unit to consider, though not necessarily the first in the chain from the transducer, is an isolation device. For whatever reason it is fairly common that a transducer and the computer which is to use it as an input are unable to share a common earth or zero point. An obvious example would be a system to monitor the fluctuations on an overhead power supply cable. The transducer is thus at many (hundreds of) kilovolts above the ground to which the microprocessor, with its console to present the results to a human operator, will be connected and an isolator will be essential. Two main methods have been used for isolation, based on magnetic and more recently optical coupling, obviating the need for a direct electrical connection and return. Figure 4.2a shows a simple magnetic coupled isolator i.e. a transformer. The main problem is that metal-cored transformers are relatively low frequency devices, limited to approximately the audio range. Ferrite cores give much higher frequency performance. Remembering that computers used to have stores with sub-microsecond cycle times which used ferrite cores, it is quite easy to see that a pulse transformer, but with more turns on its coils, can give a similar range. This is typically 3 kHz–1 MHz. Isolation of ±2.5 kV is commonly achieved, the limit being set by the proximity of the primary and secondary coils and the breakdown voltage of the insulation on the coils. However, no transformer can pass DC and so low frequency signals cannot easily be isolated by using one.

The discovery of light- and infrared-emitting diodes and complementary phototransistors

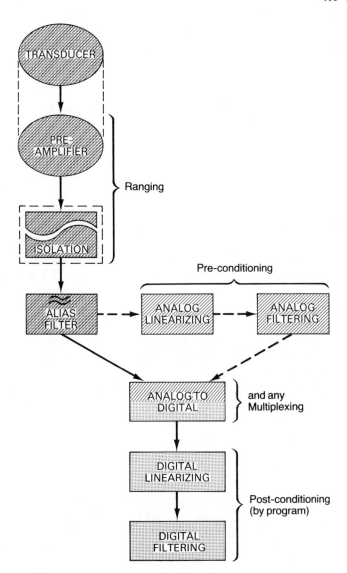

Fig. 4.1 Signal conditioning

has allowed optical isolation to become the currently dominant technique. Most devices at present use Gallium Arsenide infrared-emitting diodes (at about 850 nm) with silicon phototransistors as the initial detector as shown in Fig. 4.2b. This gives a device with a current transfer ratio (CTR = current transfer ratio/gain in percent) of only about 20% and, when fabricated in a 6-pin DIL integrated circuit, isolation of about ±2.5 kV. The circuit has a rise time of rather less than a microsecond and so gives similar performance to the pulse transformer with the important addition of low frequency and DC operation.

Many things may be done to improve on the simple opto-isolation, the first being to add a second transistor in a 'Darlington' pair configuration to give a CTR of 300% as in Fig. 4.2c. The addition of a Triac or Silicon Controlled Rectifier to the integrated circuit gives a

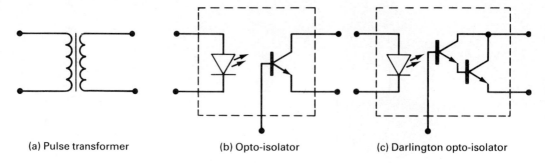

(a) Pulse transformer   (b) Opto-isolator   (c) Darlington opto-isolator

Fig. 4.2 Isolation

directly controlled, optically coupled switch, though it will be slower. Increasing the distance between source and sensor, perhaps using a fibre optic light pipe, will increase the dielectric strength and the isolation voltage to almost any value desired. For digital signals the speed may be increased above 20 MHz by using a Gallium Arsenide Phosphide emitter, low currents in the detecting transistor and a Schmidt-triggered direct logical gate included in the integrated circuit.

Optical isolators convey many other benefits, notably protecting the subsequent electronic circuits from damaging voltage transients, surges and malfunctions which may occur in harsh industrial environments, permitting widely varying input voltages via resistor/LED connection, and preventing ground loops occurring due to multiple-point earthing. These environmental problems and the solutions to them are discussed in Chapter 10.

4.2 Pre-conditioning

Prior to the ready availability of microprocessors most conditioning operations were performed by analog electronic circuits before any conversion. Pre-conditioning was the choice because the circuits were cheap compared to the cost of the minicomputers to which they were connected and because the circuits naturally operate in parallel with the processor and do not use up any of its processing time. Whilst the former reason is no longer valid the latter is still an argument in favour of pre-conditioning, particularly in extreme cases.

4.2.1 Pre-amplification (pre-ranging)

If it is necessary to cut an input voltage to get it to the correct range then a simple two-resistor potential divider can be set to give the desired attenuation. This is cheap, simple and robust. If gain is required, or conversion from current to voltage or vice versa to match an analog to digital converter input the circuits are a little more complex. Integrated circuits have flooded the analog area almost as much as the digital side. The most ubiquitous is the 'operational amplifier' which is an integrated circuit, multi-stage, DC-coupled, silicon transistor amplifier designed to have the following characteristics.

- Very high gain ($>10^6$)
- Very high input impedance
    hence low input current ($<10^{-9}$)
    and low input voltage needed.

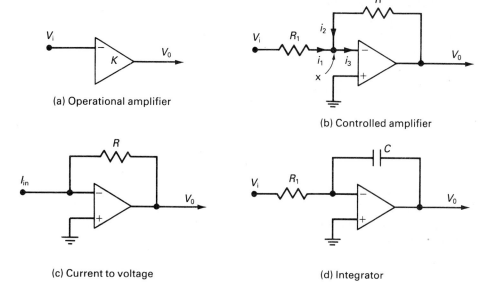

(a) Operational amplifier

(b) Controlled amplifier

(c) Current to voltage

(d) Integrator

Fig. 4.3 Pre-amplification (ranging)

- Very low output impedance
- Wide frequency response including DC
- Low drift and thermal coefficient

It is normally drawn as in Fig. 4.3a, with gain K and inversion between input and output, but is very seldom used on its own as all the characteristics are simply designed to be 'high' or 'low' and not specific. It can be used with additional components which give exactly controlled operation. A commercial **op-amp** usually has two inputs which have matched gain, with one inverting and the other not, and are summed by the operation of this differential pair input. This is followed by high gain amplifiers and a power output driver. The most common arrangement for the complete unit is with a feedback path as shown in Fig. 4.3b and this circuit will be used to explain the controlled operation. The inverting input also has the input connected to it via a resistor R_1, the non-inverting input being connected to zero volts so that it makes no contribution to the output.

By definition (design) the output voltage is related to the input to the op-amp by :
$V_0 = -K.x$ (where x is the input voltage).

$$\text{Now as } i_1 = (V_i - x)/R_1 \quad \text{and } i_2 = (V_o - x)/R$$
$$i_3 = x/Z_{in} \quad \text{and } i_1 + i_2 + i_3 = 0$$

but since the input impedance Z_{in} is very high (approximate to infinity!) then i_3 is extremely small compared with i_1 and i_2 and so may be ignored.

$$\text{Hence } \frac{(V_i - x)}{R_1} = -\frac{(V_o - x)}{R}$$

Substituting for x and rearranging gives:

$$\frac{V_i}{R_1} + \frac{V_o}{R} = \frac{V_o}{K}\left(\frac{1}{R} - \frac{1}{R_1}\right)$$

but again K is very large and so the right-hand side approximates to zero compared to either of the other terms and so

$$V_o = -V_i\left(\frac{R}{R_1}\right)$$

So amplification by known accurate amounts can be arranged simply by the ratio of two resistors, absolute magnitudes being chosen to ensure this. To emphasize the point that, as the gain is very large some voltages are minutely small and ignored, the input is often termed a 'Virtual Earth'.

$$V_o = -Kx \text{ but } K \to \infty$$

thus either $V_0 \to \infty$ which is manifestly not true
or $x \to 0$ and the input is virtually at earth.
As $K \to \infty$ and $x \to 0$ then i_3 is zero to a high degree of accuracy!

There are a number of useful twists to the basic circuit arrangements, the first of which is shown in Fig. 4.3c, for current to voltage conversion. Taking the equation for i_1 and using the virtual earth argument then:

$$V_o = -I_i R$$

It is often necessary to add or subtract a fixed (bias) voltage to the input to offset it or remove an undesirable offset voltage. A second input resistor is connected to the virtual earth point, often also called the summing junction for reasons which will become apparent and from analog computer terminology. A supply voltage or other fixed bias voltage is connected to the other end of the resistor. By the same reasoning with the new current i_4:

$$V_o = -V_i\left(\frac{R}{R_1}\right) - V_s\left(\frac{R}{R_B}\right)$$

Obviously V_s can be positive or negative and the amount of bias is set by R_B. Of course V_s could be variable and a sum of two inputs could be produced. More inputs can also be connected. There is another method of adjusting offsets using a potential divider from the output of the operational amplifier to the bias voltage supply with the offset output being taken from the potential divider centre point.

A unit gain buffer is produced by making $R = R_1$. There is no reason why we should not make

$$R = R_1 = 0$$

The feedback path is taken to the inverting input as before but the input goes to the non inverting input directly. This avoids the problem of the short-circuit between input and output which would occur otherwise.

It is also instructive to consider the circuit with a capacitor substituted for the feedback resistor. It will be necessary later on, and Fig. 4.3d shows it as an integrator. This behaviour is achieved for the following reasons:

by the virtual earth argument the input current to the óp-amp is ignored.

$$V_i - x = iR \text{ and } V_o - x = -\frac{1}{C}\int_0^t i\,dt$$

thus

$$V_o - x = -\frac{1}{RC}\int_0^t (V_i - x)\,dt$$

So again assuming x is very small the circuit is seen to be an integrator with the transfer function:

$$V_o = -\frac{1}{RC}\int_0^t V_i\,dt$$

A differentiator can easily be arranged by exchanging the input resistor and feedback capacitor.

There are many types of operational amplifier, the main differences being in the input stages, varying from the most common bipolar transistor(s) giving good general performance to the FET input type with extremely high input impedance but more susceptible to drift. It is more than likely that with improvements in fabrication technology first stage pre-ranging will be integrated with transducers in a single unit. This will allow complete calibration control during construction and give standard range signals from currently 'hard to use' transducers.

4.2.2 Linearizing

Many transducers, though otherwise most satisfactory, have non-linear amplitude responses which require straightening or 'linearizing'. If done prior to conversion an analog circuit with the inverse characteristic of the transducer is needed. This is unlikely to be easy to obtain continuously as only a few (such as logarithmic) functions are easy to produce, but reasonable approximations can be fabricated using steps of straight line segments of the form $Y = mX + c$. Figure 4.4 shows both the raw input and an approximated inverse characteristic with a circuit to produce it. No output is required until V_0 is reached by the input when the output should be

$$V_{out} = (V_{in} - V_0)\left(\frac{R}{R_0}\right)$$

until the input reaches V_1 when the output should be:

$$V_{out} = (V_{in} - V_0)\left(\frac{R}{R_0}\right) + (V_{in} - V_1)\left(\frac{R}{R_1}\right) \qquad \text{and so on} \ldots$$

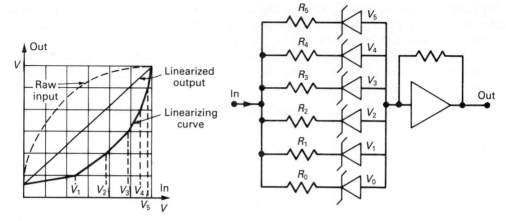

Fig. 4.4 Analog linearization

This is achieved by zener diodes of the chosen voltages which do not conduct when reverse biased until the specific voltage is reached when they avalanche and conduct, *connecting* the corresponding resistor in parallel with any others already connected. There are many other circuits which can provide similar straight line fitted curves.

4.2.3 Filtering

Apart from the requirement for an Alias Filter to remove frequencies above the double sample frequency point to prevent frequency folding there are many cases when further filtering is desirable. Only a small section of frequencies may be of interest for particular measurements such as vibration in a structure. The removal of unwanted frequencies, particularly those of superimposed noise, for example 50 (60) and 100 (120) Hertz mains hum, may be necessary to increase the accuracy of measurement. A known frequency distortion may be present in the received signal and the application of a suitable filter can correct this.

A large number of books have been produced on analog filtering and filters, particularly the common Butterworth and Chebyshev types, and both passive and active types. It is not proposed to discuss them here as, though the parallelism argument still holds as for all pre-conditioning, the flexibility and efficiency of digital filtering techniques are making them more prevalent all the time in interfacing applications.

4.3 Multiplexing, Sampling and Holding

The need for signals to be sampled quickly and held for a longer time for conversion has already been shown and circuits capable of performing these operations described. They naturally fit between any pre-conditioning and the analog to digital converter. Whilst A to D converters were expensive (and very high accuracy ones still are) there was a strong argument for sharing one converter amongst a number of inputs. The same argument applies in low-cost, low-frequency systems where a microprocessor carries out all the control operations to perform conversion, as for example in Section 15.2.

An analog multiplexer is the circuit which selects which signal shall be connected to the sample and hold, or directly to the converter in certain circumstances. Outputs can similarly be multiplexed to save on digital to analog converters by using sample/holds, or linear interpolating outputs, as described with output transducers.

An analog multiplexer consists of the same address inputs and decoding circuits as are needed for a digital selector/multiplexer to get the individual channel selecting logic signals. The difference is that instead of logic gates enabling the chosen input when given the correct select signal, an analog multiplexer uses a field effect transistor with the select signal connected to its gate input. The analog input and output are connected to the source and drain of the FET. This is very similar to the use made in the sample/hold circuit and often the multiplexer and S/H are combined by having a single 'hold' part and multiple switches to connect the various inputs for sampling.

4.4 Post-conditioning

Microprocessors are cheap and becoming faster and functionally more powerful all the time. This is the argument in favour of post-conditioning i.e. performing the conditioning operations on the digital representations of the input by algorithm in the micro. At present post-conditioning cannot cope with the highest frequencies due to the complexity and hence speed limitations of some of the algorithms needed. The great advantage of any post-conditioning is that it is completely flexible and adaptive if desired, as all parameters are set by stored constants, and functions by choice of stored (in ROM) algorithm. With early versions of 'transputers' now available and higher-speed versions coming soon, pipelined operation becomes practical and adaptive digital conditioning even more feasible.

4.4.1 Digital ranging

The accuracy of the digital data cannot be improved, of course, but it can be biased by an additive constant and ranged or scaled by multiplying or dividing by a suitable factor. The total range of numbers which can be stored in either fixed or floating point formats will have to be checked to ensure that the post-ranged values are legal. Digital ranging is simple and very fast but external ranging may be necessary as well to match input values to converter input circuits.

4.4.2 Digital linearizing

If ranging only requires a single addition and/or multiplication, digital linearizing is nearly as simple requiring a **Case** statement to provide all the necessary straight line segments.

$$\text{If } V_1 < V_{in} < V_2 \text{ then } V := m_1 V_{in} + C_1$$

Though there may be many segments and so computation of the comparisons could be slow, a look-up table can provide fast access to the gain and offset constants.

There is only one major problem which has to be considered when linearizing digitally and that is accuracy. As input is passed through any analog to digital converter its accuracy is limited to the nearest digital representation. Unfortunately when the input is non linear the slope of the input response at any point is not constant. This means that points which should be different (and would have been had the input been linear) can be given the same digital representation. No amount of calculation within the microprocessor can restore the differences as the accuracy has been lost. As an example consider a temperature input which, if linear, would move 1 K for each least significant bit of digital representation. If due to non-linearity the difference in voltage for a 1 K change is only 1/4 of a least significant bit (lsb) then three values are given the same pattern. Of course at another point on the scale more representations are being given to a 1 K change.

The only cure is to maintain accuracy by converting to a longer word, using extra bits, which can then be linearized and then normalized to the known and desired accuracy. In the example above, which is quite severe, an extra two bits would be needed. The cost is that of a more accurate A–D, the associated interface to it which may be more complex and the slower software if a double word rather than single word computation is needed.

4.4.3 Digital filtering

Any calculation can be performed on the sequence of stored samples and so algorithms to remove various frequencies are quite practical. In fact one can produce algorithms which give functions which cannot be made with analog electronics. To demonstrate how a filter can be derived using the sample data a very simple example, an *R–C* low pass filter, will be produced. This filter has an analog equivalent shown in Fig. 4.5a, though we could start by drawing any desired response on a sheet of paper.

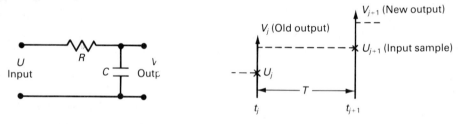

Fig. 4.5 Digital filter (sampled data)

The nature of the sampled data is shown in Fig. 4.5b and it should be noted that the input data is now piecewise constant, i.e. having taken a sample the input may be considered to have been at that value since the previous sample (or one could take it as constant until the next sample) as there is no other information about it! The sample time is *T* as before.

$$\text{From the circuit } U = iR + V \text{ but also } i = C\frac{dV}{dt}$$

$$\text{Thus } U = V + CR\frac{dV}{dt}$$

$$\text{or } \frac{dV(t)}{dt} = -\frac{V(t)}{CR} + \frac{U(t)}{CR}$$

but a general solution is known to equations of this form for values at given points.

$$\text{If } \dot{x}(t) = a.x(t) + b.y(t)$$

$$\text{then } x(t_1) = e^{a(t_1 - t_0)}.x(t_0) + \int_{t_0}^{t_1} e^{a(t_1 - \tau)}.by(\tau).d\tau$$

Thus in this case, remembering that $y(\tau)$ is piecewise constant over the samples (and sorry, τ is just a variable not an aperture time!),

$$V_{j+1} = e^{-T/RC} \cdot V_j + \int_{t_j}^{t_{j+1}} e^{-(t_{j+1} - \tau)/RC} \cdot \frac{U_{j+1}}{RC} \cdot d\tau$$

but of course $t_{j+1} = t_j + T$ where T is the sample interval!

$$V_{j+1} = e^{-T/RC} \cdot V_j + e^{-(t_j + T)/RC} \int_{t_j}^{t_{j+1}} e^{\tau/RC} \cdot \frac{U_{j+1}}{RC} \cdot d\tau$$

$$V_{j+1} = e^{-T/RC} \cdot V_j + e^{-(t_j + T)/RC} \cdot RC[e^{\tau/RC}]_{t_j}^{t_j + T} \cdot \frac{U_{j+1}}{RC}$$

$$V_{j+1} = e^{-T/RC} \cdot V_j + e^{-(t_j + T)/RC} \cdot (e^{(t_j - T)/RC} - e^{(t_j)/RC}) \cdot U_{j+1}$$

which can be expressed as a final digital equation for the new output value in terms of the previous output and current sample in two forms which are of interest.

$$V_{j+1} = U_{j+1} + e^{-T/RC}(V_j - U_{j+1}) \qquad \dots\ \text{form A}$$

Form A requires a constant (for chosen R, C and sample time) which is multiplied by the difference between the previous output and current input sample, with an addition to give the new output. Although this is the fastest algorithm (there are similar rearrangements of course) it has one serious problem. The accuracy will be limited by the difference term when the values are close due to the finite word length. Another rearrangement is.

$$V_{j+1} = U_{j+1}(1 - e^{-T/RC}) + V_j(e^{-T/RC}) \qquad \dots\ \text{form B}$$

This requires two multiplications and so is slower for most microprocessors, but it is accurate. It is common practice to draw such equations as transfer functions in a similar way as is done for control systems and *form B* is shown in Fig. 4.6 in this way.

A slightly different response would be obtained if the calculation was done assuming the input sample was constant **until** the next sample rather than **from** the previous sample, using U_j not U_{j+1}. The difference is that between the Butterworth and Chebyshev forms.

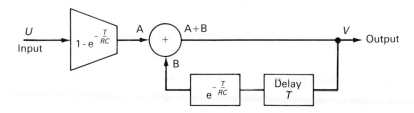

Fig. 4.6 Digital low pass (*CR*) filter

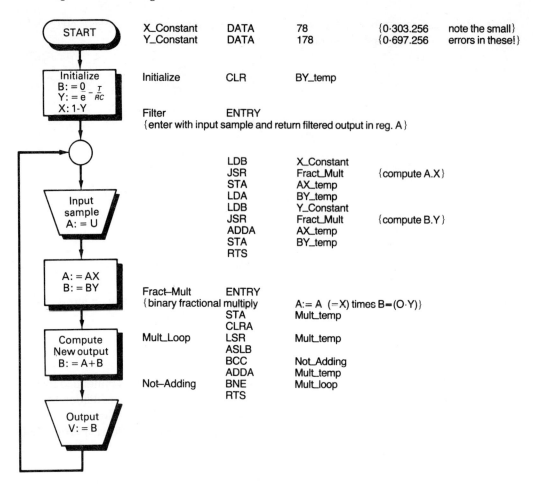

Fig. 4.7 Digital filter algorithm and program

The complete algorithm and program (given in simple Motorola 6800/2 series code) are shown in Fig. 4.7. For most simple micros, not having a multiply instruction, the fractional multiply algorithm is quite quick. One can compute all possible values which could result from a multiplication (for 8-bit accuracy only 256 results) and store them in a table using direct look-up to 'compute' the result. This is faster than the algorithm shown which gives the time responses to step and triangle waves shown in Fig. 4.8 and would allow a much shorter sample time. Obviously any order of filter could be produced by an entirely similar process to give filters of any analog characteristic though it is better to cascade first or second order filter algorithms (pipeline!) rather than compute a single higher order form directly because of finite word length effects. This is because the poles of a transfer function are shifted due to word length errors from their exact points and this reduces stability.

The minimum necessary word length can be shown to be both proportional to the order of the function, hence cascade lower orders to make a higher order, and also proportional to $\log (1/T)$, so the sampling frequency should be minimized! Any digital filter can be worked out from a knowledge of the locations of poles and zeros in the continuous plane (the

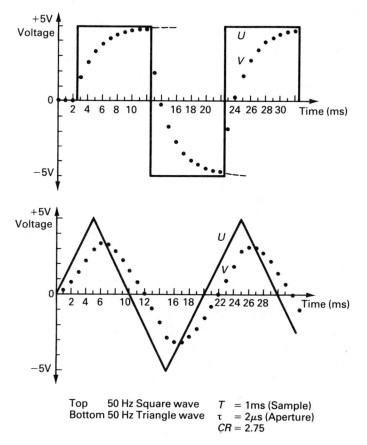

Top 50 Hz Square wave T = 1ms (Sample)
Bottom 50 Hz Triangle wave τ = 2μs (Aperture)
                                          CR = 2.75

Fig. 4.8 Response of digital low pass filter

S-plane) of the desired characteristic to be seen outside the micro. The equivalent poles and zeros in the sampled plane (the Z-plane, used inside the micro) can be mapped using the bilinear or matched-Z transforms from the continuous plane. A completely artificial digital filter, with no analog electronic equivalent, can be provided by starting from any desired characteristic.

A final cautionary note on digital filters with lead as opposed to lag characteristics, i.e. **high-pass** types. As differentiation is involved the algorithm will contain differences which cannot be avoided. For example swapping the C and R in Fig. 4.5a to give a high-pass filter results in a transfer function of:

$$V_{j+1} = U_{j+1} - U_j - V_j(e^{-T/RC}) \qquad \dots \text{ high-pass } RC$$

To retain the same accuracy as the low-pass case, a microprocessor with 16-bit operations (rather than 8-bit) would be required.

5

Signal Conversion: Digital to Analog

Computers only operate on digital values and so any analog signal, like most signals from the real world, must be converted to the nearest digital representation to the desired accuracy (plus or minus one half of the least significant bit). The bounded infinite set of analog values is **quantized** to a discrete set of digital values by comparing the unknown input with known digital equivalent values to find the nearest match. Thus, before considering analog to digital conversion we should look at digital to analog converters which could provide the known digital equivalent values as output.

D–A converters either provide voltage or current outputs but for greatest speed they switch currents internally. Current steering is faster because the reference current is not switched on or off and the only significant voltage changes are the required ones. The basic technique is to apply the digital pattern via a set of switches on to a precision resistor network to which an accurate reference voltage source is connected. The output is the sum of the selected currents and is output either as the current or converted by an operational amplifier to voltage. There is another, slower technique based on pulse width output which is described later.

5.1 Simple Weighted Network D-A

Conceptually, the simplest technique for a D–A is to have binary-weighted resistors powered by a fixed reference voltage. The digital pattern to have its equivalent voltage output is connected to switch in the currents generated for each bit. These currents are summed by the output operational amplifier and either produce a voltage or current output. This is shown in Fig. 5.1a in concept. Typical details of the switching circuit are shown in Fig. 5.1b. Each transistor/resistor pair forms a precise current source which may be switched in

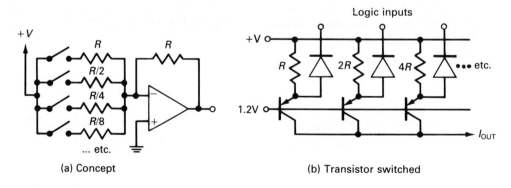

(a) Concept   (b) Transistor switched

Fig. 5.1 Simple weighted digital to analog converter

by the logic inputs from the digital word. The resistor R supplies a current I, the resistor $2R$ supplies a current of $I/2$, $4R$ gives $I/4$, etc. down to the least significant bit and the currents are summed.

Unfortunately, this simple system, though it can be very fast, does not actually work in practice except for short word lengths. Assume an eight-bit word, then resistors R . . . $R/128$ are required. If the resistor R is made with an accuracy of 1% then the error in current caused by this is greater than the contribution of the least significant bit. If resistors are made with greater accuracy, perhaps by laser trimming, then we simply have to extend the word to reach a circuit which fails due to the error. It is impracticable to make sets of resistors have large ratios accurately. Happily a simple alternative is available which forms the basis of the vast majority of D–A converters.

5.2 *R-2R* Weighted D-A

Looking at Fig. 5.2, we can see the complete circuit of an R-$2R$ weighted D–A converter with its network, sometimes called a ladder network for obvious reasons. A small part of it is expanded to help in the explanation of its operation. Consider looking at junction X (Fig. 5.2), any typical junction on the ladder, and, if you can, put your eye at position 'c' and observe. To the right you would 'see' the $2R$ resistor running vertically in parallel with everything beyond it, i.e. what you could see from 'b'. Similarly viewing from 'd' to the left, you would see $2R$ in parallel with the rest of the circuit to the left, i.e. what you could see from 'a'. Exactly the same will hold for all other junctions until either end of the ladder is reached. At the left-hand end of the ladder is a $2R$ resistor which represents the rest of the circuit to the left for the first junction (from 'a'). Similarly at the extreme right of the ladder is a $2R$ resistor which represents the rest of the circuit to the right of the last junction (from 'b'). Remember that the input to the operational amplifier to which this resistor is connected is a **virtual earth** and so may be considered to be connected to earth, like that at the left-hand end, to a good approximation. Also remember that a voltage reference must be capable of remaining at an unaltered voltage no matter what current is taken. This means that its impedance to the earth return must approximate to zero. All the factors controlling the operation of the circuit are now to hand.

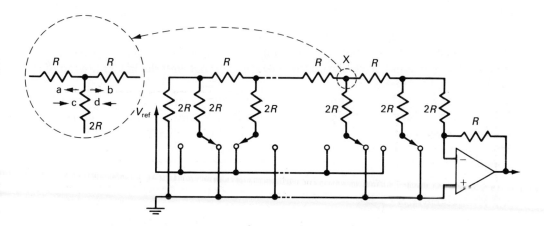

Fig. 5.2 *R-2R* weighted digital to analog converter

Thus, starting from the right-hand end there is $2R$ in parallel with $2R$ giving R which is in series with an R to point X (view 'b') giving $2R$. This is in parallel with a further $2R$ giving R viewed from point 'c'. The circuit is symmetric so it is simple to deduce the other two views giving:

> View from 'a' see R in series with R giving $2R$
> View from 'b' see R in series with R giving $2R$
> View from 'c' see $2R$ in parallel with 'b' $2R$ giving R
> View from 'd' see $2R$ in parallel with 'a' $2R$ giving R

Now starting with the left-hand junction, which will be the least significant bit, if its switch is set to ground then no current is introduced into the ladder. If, on the other hand, the switch is set to V_{ref} then a current will flow through the $2R$ resistor to the junction. There the current divides, 'seeing' paths equivalent to 'a' and 'b'. To the left is $2R$ to ground. To the right is R in series with the rest of the circuit in that direction, a view 'c' on the right giving another R, so this too is $2R$ to ground. The current thus divides equally, half going to the right and half to the left. Hence, if V_{ref} is switched in a current of $V_{ref}/3R$ passes down the resistor to the junction and so $V_{ref}/6R$ goes to the right to the next junction. Here assume the switch is set to earth, then the current divides between the $2R$ to the switch and the total resistance of view 'b' at this junction—also $2R$.

So, to sum up, at any junction the current flowing into the junction from the left divides equally leaving half to run on to the right. A current of $V_{ref}/6R$ or a current of **zero** is added to this depending on the position of the switch. This summed current then flows to the next junction when the same occurs again, a division by two and addition of either 'zero' or 'one' current. For Fig. 5.2 with the LSB to the left and the digital number 0010 set the output current is

$$\left(\left(\left(\left(\frac{0}{2}\right)+\frac{V_{ref}}{6R}\right)/2+0\right)/2+0\right)=\frac{V_{ref}}{24R}$$

If the switch below X is set the other way then the output current is

$$\left(\left(\frac{V_{ref}}{12R}+\frac{V_{ref}}{6R}\right)/2\right)=\frac{V_{ref}}{8R}$$

This current can be output or is converted to a voltage by the operational amplifier which is arranged to have unit gain as shown. Real transistor switches are used of course instead of those shown, and could be similar to those of Fig. 5.1b but with the ladder instead of the individual weighted resistors.

The R-$2R$ weighted D/A is simple to construct with high accuracy and is commonly found in 8, 10, 12 and 16-bit word lengths. The basic circuit shown only provides outputs from zero to V_{ref} and would require an additional offset to give both polarities of output. Also note that the accuracy is largely determined by the accuracy of V_{ref} once the resistors are accurate enough.

5.3 Multiplying D-As

A simple extension to the circuit permits multiquadrant operation. This also allows a very useful extension to the range of function of D-A circuits. Firstly there is no reason why the

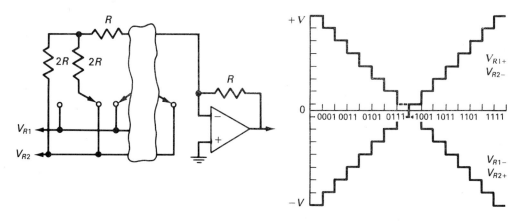

Fig. 5.3 Multiplying and multiquadrant D-A converters

switches should not operate between two voltage references (say V_{R1} and V_{R2}) rather than V_{ref} and earth (zero volts reference). Secondly there is no reason why the voltage reference(s) should be of fixed polarity, and thirdly there is no reason why the references need be fixed at all! Figure 5.3 shows both the connection of the basic R-$2R$ converter to two references and the outputs which can be generated from fixed references of varying polarity (V_{R1} positive $= +$ etc.) showing multiquadrant operation.

A positive reference and a negative digital number or a negative reference and a positive digital value give negative outputs. The fourth quadrant is generated by two negative values.

The output of a multiplying D-A is the product of the digital value set to the switches and the instantaneous value of the excitation voltage (the 'reference'). There are a multitude of uses for this function, some of the more obvious applications being: digital control of audio signal intensity, digital phase shifting circuits, control of raster scanning for cathode ray tube displays, complex function generators, digital modulation and demodulation, and automatic calibration systems.

Digital to analog converters using the principles described are made in various word lengths from 6 to 16 bits and with settling times from a few microseconds down to a few nanoseconds in units available commercially.

5.4 Stochastic D-A

Though the vast majority of converters use resistor ladders, there is another technique which though slower is significantly lower in cost and integrates into digital techniques more easily. The trick is to use an intermediate variable between the digital value and the analog voltage. Microprocessors and digital circuits in general are excellent at counting and timing so time is the intermediate chosen. The digital value is converted to a time ratio which is output as a pulse width with:

$$\frac{V_{out}}{V_{max}} = \frac{T_{on}}{(T_{on} + T_{off})}$$

The pulses are output as V_{max} for time T_{on} followed by zero for time T_{off} in a continuously repeated cycle until a new V_{out} is required. The pulse train is then integrated (smoothed) by an external integrating circuit such as that of Fig. 4.3d, or even a simple capacitor/resistor low pass filter.

The output from this pulse train when fed to a perfect integrator for infinite time should be:

$$V = \int_0^\infty \frac{T_{on}}{(T_{on} + T_{off})} \cdot V_{max} = V_{out}$$

The long-term average of the pulse train is obviously the correct mean V_{out} value but there is a problem as the integration for time intervals of less than $T_{on} + T_{off}$ will be incorrect. The problem can be visualized by thinking of the capacitor 'store' of the integrator being charged up for time T_{on} and then discharging for time T_{off}. The capacitor is then topped up, discharged, etc. and so the output fluctuates about the desired value. The solution to this is

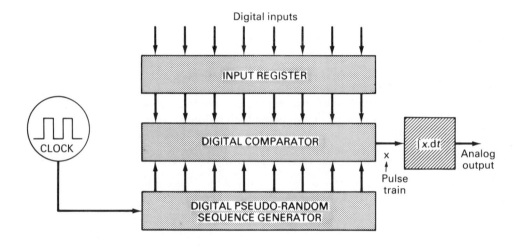

Fig. 5.4 Stochastic digital to analog converter

to distribute the T_{on} into many more, shorter ON times, and distribute the T_{off} into a similar number of OFF times leaving the total on and off in a cycle unaltered but spreading them evenly through the cycle.

A simple method to achieve this distribution is to use a Pseudo Random Binary Number generator (hence the name stochastic) with the circuit of the converter as shown in Fig. 5.4. Note that all parts except the integrator could be performed by software instead of explicit hardware components. The required digital output value is compared with the present value of the PRBN and if the digital output is greater than this, then V_{max} is output for the duration of that minor cycle, otherwise there is a zero output. Figure 5.5 shows the waveforms generated firstly from the PRBN for **two** full cycles each of **eight** minor cycles for a very simple **3-bit** converter, and secondly the output at 'X' (the input to the integrator) for each of the eight possible values of output. Each is shown for two major cycles and the improved distribution is apparent for values of two to six.

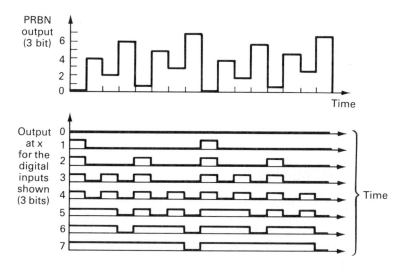

Fig. 5.5 Waveforms produced in a stochastic converter

Of course, there are now 'minor cycles' but the same timing accuracy is required for a given digital to analog accuracy as for a simple pulse width cycle. The drawback of stochastic converters is, of course, their speed. To get an accuracy of ten bits there must be 1024 possible time divisions for each major pulse. Those are the times when the single large pulse turns off or for each smaller minor cycle to be either on or off in the stochastic version. If each of these divisions is, say, one microsecond (i.e. using a 1 MHz) then a major cycle is more than one millisecond. The integrator must integrate a number of such cycles before the output will settle and with a single pole integrator, with a time constant of four and a half milliseconds, the settling time will not be much less than thirty milliseconds. So, assuming four outputs per cycle, the maximum frequency of output is less than 10 Hz. Even this is based on the assumption of a minor cycle every microsecond. If the digital parts are replaced by software then with the generation of the PRBN cycle, the comparison and output instructions and a fast microprocessor with little else to do we will be lucky to exceed a 1 Hz wave output at ten bits accuracy. Stochastic converters are, however, very useful for low-speed, low-cost outputs where the micro can be used to do the work.

5.5 Converter Coding

In the previous discussion of conversion, binary encoding was assumed but this is not universally used. Converters can provide unipolar (single sign) operation in binary or binary coded decimal (BCD) forms, or bipolar operation in either of these in sign and magnitude form. Bipolar operation will also be supported by complement coding, either one's or two's complement or by offset binary code. Any of these codes can be driven by a positive or negative logic convention.

The simplest coding is obviously straight binary for unipolar operation and offset binary for bipolar operation. These directly match both $R/2R$ or weighted current configurations. Note that offset binary has the most negative value represented by all zeros, zero volts by a one followed by all zeros and the most positive value represented by all ones. Simple conversion from stored data applies for this. Two's complement coding might be preferred as it fits the coding of most microprocessors directly. It can be arranged by having the most significant bit (sign bit) supply to the opposite input (inverting) of the output amplifier. This avoids most current 'glitches' at the sign change over. Most other techniques cause problems and so two's complement is not as popular as might be expected for D-As. Both one's complement and sign and magnitude binary codes have the problem of two zero representations, both plus zero and minus zero. Sign and magnitude can be arranged by switching the reference, but this is not commonly used. BCD encoding requires an additional decoding stage and does not fit in easily with most microprocessors.

5.6 Converter Accuracy

Unlike acquiring an analog value and converting it to a digital form there *should* be no error when a digital value is converted to analog form as one of a small set of discrete values is selected. There are many possible inaccuracies however and one can rely on manufacturers quoting figures to put their products in the best possible light. Figure 5.6 shows the ideal output and the various errors. All errors *should* be quoted over the full range, for example as a percentage of full range rather than the suspect and less stringent percentage of reading or value.

The accuracy criteria and errors for digital to analog converters are similar to those covered for transducers in Section 2.2. The **range** or **full range** is the maximum value of analog voltage which the D-A can output, and should include its ability to handle multi-quadrant operations. Some converters handle a skewed range, not covering the same range

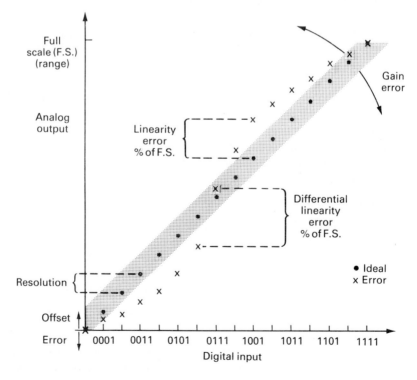

Fig. 5.6 Digital to analog converter errors

for negative as for positive values. The **resolution** is determined simply by the number of bits used by the converter and specifies the minimum value difference which can be output. Faults in the uniformity of the converter's output are specified by **monotonicity** and **linearity** errors.

A converter is monotonic if for all increases in its digital input the analog output increases or remains the same but does not decrease. Similarly for all decreases in the digital input the analog output decreases or remains the same but does not increase. Obviously the same definitions hold if the output increases for decreases in input or vice versa as long as there are no reversals in direction. Non-monotonic behaviour is aggravated by temperature coefficient effects and can be caused by switching time defects causing bits to switch at different times. The effect can be serious for on-line control as any faulty change in direction of output may reverse the direction of control.

The linearity of a converter, sometimes called its relative accuracy, is measured as the maximum deviation of the analog output from a straight line joining the digital end points of the range. This end point definition may not be achieved in practice if there are gain or offset errors between values within the range. A given linearity error in a digital to analog converter output cannot be corrected subsequently, and the only solution is to use a converter with better resolution and designed with better linearity.

A better measure of the linearity errors inherent in a given converter design is **differential linearity**. This is defined as the maximum deviation in the analog difference between two adjacent output codes from the ideal value, i.e. the full range divided by the number of steps, or $V_{max}/2^n$. It is easy to see from Fig. 5.6 that this can give much worse errors than the 'integral linearity' described before. It is worth noting that a converter may remain monotonic even though it has differential linearity errors of greater than one least significant bit, though if the differential linearity error is less than ± 1 lsb then the output must be monotonic. Also there can be no missing or skipped codes if this latter condition applies.

The worst differential linearity errors for the common R-$2R$ converters occur at the major code changes, for example when changing from 1000 . . . to 0111 . . . etc. From the description of operation in Section 5.2 it was apparent that the contribution of each low-order bit was divided by two at each junction it passed, contributing to any error. Code 1000 . . . only has an error associated with inaccuracies in the rightmost R-$2R$ pair of Fig. 5.2. By contrast 0111 . . . has contributions to the error from all the other R-$2R$ pairs to the left and hence the transition between these adjacent values gives rise to the worst differential error. There is a further problem at such major code changes but this is discussed further on with timing problems.

Of course, there are circumstances when a purely linear relationship is **not** desired, the most common non-linear converters being **logarithmic** in their transfer characteristics. If a linear characteristic is used then there is a greater relative error as a fraction of the signal when the signal is small. If the error, due to any cause, is to be kept as a constant percentage of the signal then more steps, closer together, must be allocated closer to the origin and fewer further away. This is exactly the response achieved by a logarithmic weighted D-A and Fig. 5.7 shows the characteristics for simple 3 bit D-A and its inverse A-D converter.

A further two errors shown on Fig. 5.6 are the **gain** and **offset** errors though these can be corrected during system design or whilst in use. The offset error is defined as the deviation in analog output from zero when the digital code for zero is selected. This is the amount by which the graph of the output fails to pass through the origin referred to the analog output axis. It is caused by inaccuracy in the output amplifier, possibly temperature related, and in the resistors controlling it, but is adjustable to zero with most D-As by a trim control. The gain error, sometimes called the **scale** error, is defined as the percentage difference in slope between the actual transfer function and the ideal straight line. It is caused by inaccuracies in the reference voltage, the output amplifier and the ratio of its gain control resistors. It is similarly adjustable to give the correct gain (slope) by a trim control input on most D-A converters.

Most of the parameters of D-A converters are subject to variation if the operating temperature is altered. The temperature coefficients are quoted relative to a standard temperature of 25°C though sometimes relative to a slightly different base. Linearity errors are quoted in parts per million of full scale per degree. This is the most serious temperature-related error as it cannot be removed by calibration. Differential linearity temperature coefficient, quoted in ppm/°C, is even more serious as operating away from the standard temperature may take a converter from better than ±1/2 lsb error to greater than 1 lsb and

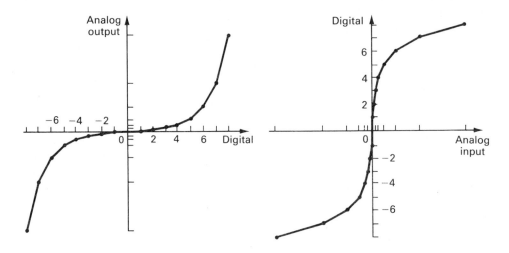

Fig. 5.7 Deliberately non-linear D-A and A-D

maybe to non-monotonic behaviour. For example, a 12-bit converter has a resolution of approximately ±120 ppm but to guarantee that differential linearity error does not result in non-monotonic behaviour when operating at 30° above standard temperature a DLTC of better than 4 ppm/°C would be required.

Gain temperature coefficients are quoted in ppm/°C or by a 'butterfly' graph showing the different limits operating over different temperature ranges. Offset temperature coefficients are either quoted in ppm of full scale/°C or more commonly in microvolts per degree centigrade. Both of these errors can cause greater than 1 lsb error if not calibrated as commercial gain or offset TCs are often 5 to 50 ppm/°C.

The last static operating error is sensitivity to changes in power supply. This is quoted as a percentage change in analog output for each 1% change in power supply. It should be well below the resolution limit for a 5% supply change for any good converter, but it can add to all the other error contributions to give a significant worst case error.

The most serious drawback of any given converter, however, will be its speed of operation. This has the same components as the response time of transducers shown in Fig. 2.2. The delay plus the slewing and any overshoot sum to the **settling time**. This is the time taken from the application of a full range step to the switches until the output enters and remains within the resolution of the chosen output. This does not include the time taken to load the switches with the new value, termed the **switching time**, which is important in very high-speed units. If an eight-bit processor is linked to a 12-bit converter then two loads are needed to set a new value and double buffering is necessary to ensure that no false 'half loaded' value is output.

It is possible for the settling time to be greatest not for a full scale swing, but for a single lsb change at the major carry, i.e. when a single bit added causes all the other bits to change. Settling times vary from less than ten nanoseconds for eight-bit current steered converters using emitter-coupled logic interfaces through 500 nanosecond twelve-bit *R-2R* layouts to the stochastic operation at many milliseconds.

Finally, the switches of resistor ladder D-A converters cannot operate absolutely simultaneously and so whenever two or more switches must change to give a new value there will be a transient over supply or inadequacy of current. This will be worst at the major carry 0111 . . . to 1000 . . . etc. There is also a contribution to such **'glitches'** by capacitive coupling and leakage through switches and from components located nearby, particularly in monolithic integrated circuits. Glitch amplitudes can be quite considerable and unless the output is already strobed such as in visual display systems the only solution is to add strobing using a sample/hold amplifier. This ensures that the settling time has passed and all transients died away with the previous value **held** before the new value is let through in **sample** mode.

To sum up, the fact that a converter is a 12-bit converter in no way guarantees that its output is accurate to 12 bits. In many cases the least significant two or more bits are simply swamped by the errors described due to the chosen operating conditions. So to choose a converter. Firstly choose the speed of operation needed. The three types: current steered, *R-2R* or stochastic give settling times of 10 ns or less, 50 ns and up, or in the millisecond range respectively. Secondly choose the resolution required: this can range from 6 to 16 bits and even up to 20 for use in standards laboratories. The current steered, binary-weighted mode of operation cannot operate at long word lengths as previously explained and the longer the word the slower is the possible operation of a stochastic converter. Thirdly decide on the operational factors such as temperature range, power supply sensitivity, output glitch prevention and which calibration procedures are acceptable. Review all the other errors described in this section to ensure adequate operation. Fourthly choose the type and range of output signal required, deciding on current or voltage output. The standard full scale ranges of commercial units are ±2.5V, ±5V or ±10V for bipolar units and 0 to +5V, 0 to +10V, 0–2 mA or 0–5 mA for unipolar units. The fifth consideration is the kind of reference required, either internal or external, fixed or variable as in multiplying D-A converters. If the reference is to be variable then in which quadrants must it operate? Finally consider the coding required to be presented to the digital inputs, normally straight or offset binary, and the interface by which the inputs connect. The interfacing of converters to micros is covered in more detail in Section 6.5.

6

Signal Conversion: Analog to Digital

Most operations involving computers have two extremes for the performance of their algorithms. They can be fully parallel or fully serial. There are also many intermediate forms, in many cases, combining serial and parallel computation. For example, multiplication can be performed fully serially by repetitive addition or by shifting (multiplication by two) and addition—combining serial and parallel, or even by a table lookup of all possible values giving a fully parallel and very fast algorithm. Analog to digital conversion is no exception; the fastest converters are fully parallel, the cheapest fully serial, with a plethora of types in between. The range of speeds available has improved dramatically since the early decades of computing when conversions took, at best, 20 microseconds—now the same number of nanoseconds will suffice.

An A-D converter using a serial algorithm needs the ability to generate the complete set of discrete analog values one at a time. The unknown input is then compared with these, in some order, to determine which is the nearest approximation. Hence, a D-A converter (to produce the values) and a comparator are the only necessary hardware elements. A parallel converter would check all possible values at the same time and thus needs considerably more circuitry.

6.1 Parallel Converters

To convert an unknown voltage to its nearest digital representation in the fastest way a **parallel** or **flash** technique is used. This has one comparator and one analog voltage reference fixed equal to the digital representation for each possible digital number. Thus for a six-bit converter sixty-three comparators and sixty-three references are required. This limits the word length possible with this technique to that which integrated circuit technology can pack onto a chip. It can be seen from Fig. 6.1 that the generation of the discrete fixed reference is easy and more accurate than the R-$2R$ ladders of D-A converters. A three-bit flash converter is shown to simplify the drawing. The resistor chain which develops all the intermediate references from the master voltage reference (V_{ref}) is connected to each comparator. The output from each comparator is connected to a $2^n \rightarrow n$-bit encoder which takes the output bits 0/1 from each and encodes the result to the single binary value. Cheap flash converters are commercially available in four-, six- and eight-bit configurations with ten bits now becoming practical. Conversion times of less than ten nanoseconds are now available with short word lengths and, by being designed with better than $\mp 1/4$ lsb linearity, converters can be operated in pairs to give an extra bit of resolution. This combination shows how a 'nearly parallel' operation can achieve very high speed by feed forward methods which are described in Section 6.3.

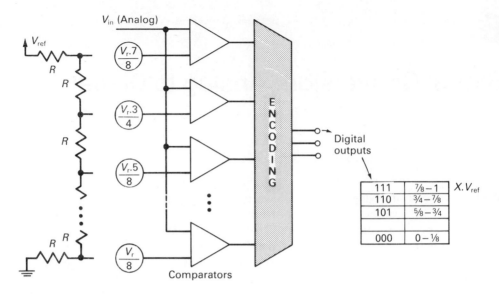

Fig. 6.1 Parallel analog to digital converter

6.2 Counting and Tracking Converters

At the opposite end of the parallel/serial spectrum come counting converters. These use a single D-A converter to provide all the reference voltages and step through trying each in turn until the nearest is found. This is a feedback approach. Only a single comparator is needed to compare the unknown input with the selected trial output from the D-A.

The simplest, but also the slowest, is the counting converter shown in Fig. 6.2. The basic hardware is shown with its operation and, as with all the other analog to digital converters in this chapter, their control and register functions can be performed by software in a microprocessor.

The n-bit counter is reset to zero to start a conversion. The output of this register is connected to the D-A converter so the smallest voltage is output and compared with the unknown input. A clock steps the register trying all values in turn until the comparator output changes. This signal disables the counter clock indicating that the digital equivalent of the input is held in the register. This gives a variable conversion time (as shown) but the maximum time is 2^n times the clock time. This could be reduced by various tricks such as splitting the register and counting only the top half (most significant bits) first. Once the comparator output changes then the top half of the register is decremented by one. The bottom half of the register is then stepped up from zero until the comparator output again changes to indicate that the complete correct value is held. The basic counting technique is for comparison purposes only, however, and would never be used, but it is the basis for the much faster tracking circuit.

As an aside it should be noted that the clock time determines the ultimate speed and it is fixed by the speed of the logic to step the counter, the settling time of the D-A converter (as in Section 5.6) and the comparator delay. The analog parts can be speeded up by switching and comparing currents rather than voltages as previously described. No operational amplifier is required to convert the summed current of the D-A to a voltage prior to

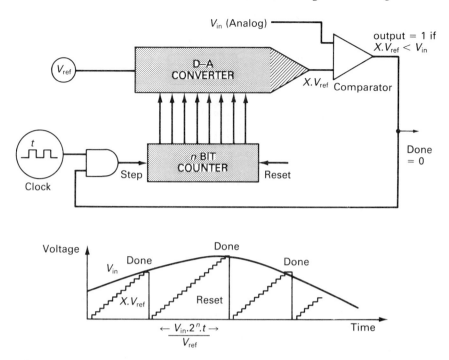

Fig. 6.2 Counter analog to digital converter

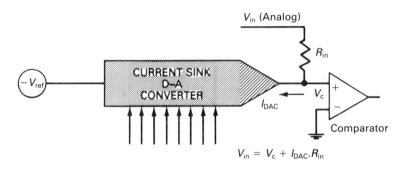

Fig. 6.3 Current comparison input and DAC

comparison. The connection of a current output D-A for this operation is shown in Fig. 6.3 and is common practice.

The individual step of one bit change to find the input is very fast, it is just the number of steps which makes the converter so slow. If one could arrange to need only **one** step to find the unknown input the converter would be only marginally slower than the flash converter yet give greater accuracy.

Having used a counting technique **once** to find the unknown input, if we take a further sample soon enough then we know within plus or minus one bit what the new input will be. Thus if, instead of resetting the counter, we start it from its previous value, we must take the minimum time to find the new input. This is the principle of the **tracking** (or servo) converters, and the basic circuit and operation are shown in Fig. 6.4. A similar D-A

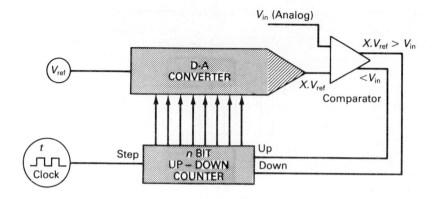

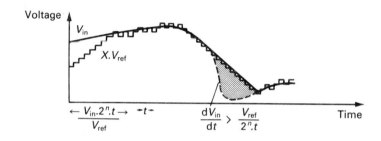

Fig. 6.4 Tracking analog to digital converter

converter is used but the counter must be able to step up or down. The direction is determined by the comparator with some logic to enable the clock to count up or down, but otherwise the circuit is similar to the previous type. The operation starts in an identical fashion though as the circuit 'counts' to find the initial unknown input. This is signalled by the first reverse of the direction of counting. Then the counter either steps up or down depending on the result of the previous comparison. The register always holds the current approximation in digital form to the unknown analog input.

Of course there is a snag! If the input signal slews faster than the tracking converter can step then the signal is 'lost' and the counting has to be restarted and run until the signal is reacquired. During this time the register holds an incorrect value. To avoid this fault the condition:

$$\frac{dV_{in}}{dt} < \frac{V_{ref}}{2^n.t} \text{ must be maintained}$$

If, however, this cannot be guaranteed we can still ensure that the converter is useful if a little extra circuitry is added. If the tracking fails to catch up in two steps then instead of stepping the least significant bit the counter changes to step in increments of twice the size (the next least significant bit). If this fails then the next bit up is stepped until the signal is re-acquired when steps revert to the smallest increment. The most significant bits of the value in the register are always correct. Its accuracy (in bits) reduces when the input slews too fast.

This is, in fact, the next fastest type of converter to the flash converter if only a single input is connected to it. The time for a step to a new (correct) value or the loop slew rate is equal to the sum of the comparator response plus the D-A settling time plus the counter propagation delay. It cannot be used to advantage with multiplexed inputs as the benefit of tracking is lost. If a direct current input is connected to a tracking converter the valve in the register will oscillate (alternate) between two adjacent values, the difference being the quantization error.

If a multiplexed set of inputs is to be connected to a single converter then tracking will not work, unless the last value for each channel can be stored and reloaded, hence another approach is required.

6.3 Successive Approximation Converters

The counting algorithm is faster for inputs nearer zero. For a general-purpose converter all possible inputs are equally likely and so an algorithm with no similar 'bias' is desirable. Such an approach is common in other computing applications and gives rise to binary chop searching which is the basis of successive approximation converters. The advantages are the simplicity of the algorithm and the hardware, the speed (as only n steps are taken for an n-bit word) and the consistency as the time is constant for any conversion.

A conversion starts by comparing the unknown input with an output from the D-A equal to *half* the maximum range of possible inputs. If the result indicates the unknown input is greater than the D-A output then the next test is against *three quarters* of the maximum. If the test indicates it is not greater then the next test is against *one quarter* of the maximum. For each bit in the word the test is on half of the range of the previous test.

i.e. the result is $X_8 X_7 X_6 X_5 X_4 X_3 X_2 X_1$ where $X_n = 0$ if $V_{in} < V_{out}$

or $X_n = 1$ if $V_{in} > V_{out}$

where V_{in} is the (constant) unknown input voltage
and V_{out} is the cumulative value of $X_n X_{n-1} \ldots$. down to the currently tested bit.

Thus the conversion proceeds by comparison with a series of trial values and the result of each gives both one more bit of accuracy to include in the final value, and the next trial value. Figure 6.5 shows the basic hardware and operation of a successive approximation converter. A shift register is cleared and has a '1' set into its most significant bit. The result register starts cleared and the sum ('OR' is perfectly satisfactory) of the two registers is output to the D-A. The output of the comparator controls the action, either copying the shift register bit (again OR-ing) into the result register, or not copying it, and then shifting the bit one place down to perform the next test. When the shift register empties (by shifting the 'test' bit off the end), the conversion is complete. The result register holds the digital approximation to the input. The successive approximation technique is the basis of all fast, general-purpose converters and is applicable to both multiplexed and non-multiplexed inputs directly.

The successive approximation converters are not as fast as the parallel type yet can give greater accuracy. Combining the various techniques can yield benefits. If a partial accuracy conversion is done by a simple flash procedure, the result may be used to set a base for a smaller range conversion by another, or the same, method. These are **feed forward**

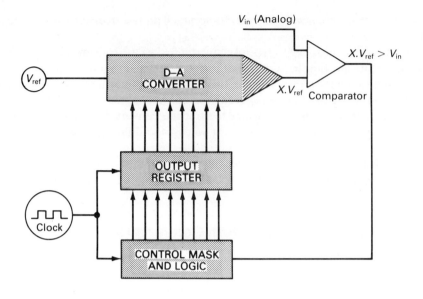

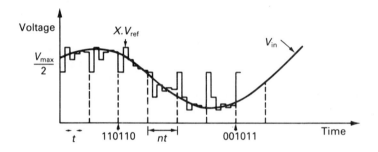

Fig. 6.5 Successive approximation A-D converter

techniques as the result of one conversion (giving, say, the most significant four bits) is fed to a subsequent stage for the next four and so on. This can obviously be faster, but each stage must be as accurate as the least significant bit. One way of achieving all this is to have the first A-D set its result to a D-A which alters the reference for the next A-D stage which operates on a reduced range (a sixteenth in this case) and so on.

6.4 Intermediate Variable Converters

The problem with the converters discussed so far is that they rely on the stability and accuracy of a number of components for their own resolution and accuracy. Flash converters need 2^n accurate reference voltages and comparators, and feedback types need a digital to analog converter based on a resistor ladder. Such converters are limited to twelve to fourteen bit accuracy. If a lower speed of operation is acceptable then a much cheaper approach which is capable of very high accuracy is available.

The use of an intermediate variable is the key. First converting the unknown voltage into some other form and then converting that to a digital representation permits more accurate resolution of the unknown input if the two stages do not rely on component stability. As with the stochastic digital to analog converter, time is chosen as the intermediate variable. This is because digital electronic circuits are able to measure time intervals so accurately.

The simplest circuit is a **single slope** integrator. A single comparator is used with the unknown input connected to one side. The other is connected to the output of an integrator with its input connected to a fixed reference voltage. This gives a precision ramp generator which is reset to zero volts at the start of each conversion. A counter is also cleared and allowed to count up until the precision ramp output is found to be greater than the unknown input by the comparator. Once the comparator switches the counter is stopped and the time measured is proportional to the input. This approach is too simple, though, as the ramp depends for **its** accuracy on the stability of the capacitor in the integrator. If the clock varies in frequency over a period of time the same input will be converted to different values. And finally, as the conversion time varies with input voltage, specific noise frequencies are not easily rejected.

6.4.1 Dual-slope converters

To remove these errors a **dual-slope** technique is preferable. Figure 6.6 shows both the basic circuit and its operating principle. The same basic counter, comparator and integrator are used and the output of the integrator is still connected to one input of the comparator. The other input to the comparator is connected to a fixed point—zero volts in the figure. The input to the integrator can be switched to either the fixed reference (now negative to give a decreasing ramp) or to the buffered unknown input. Thus two 'ramps' can be created, one determined by the unknown input and the other by the fixed reference.

A conversion commences with the switch set to connect the negative reference $(-V_{ref})$ to the integrator thus causing it to ramp down to the starting-point. To begin a conversion the switch is set to connect the unknown input to the integrator which ramps up at a rate determined by the unknown voltage, as shown in Figure 6.6. When the output of the integrator crosses zero, as indicated by the comparator (which has one input connected to zero) switching its output, the counter starts counting. After a **fixed** count 'T_1' the switch is set back to the negative reference, the counter is cleared and counting restarts. When the comparator again indicates the integrator output passing zero, following the downward ramp towards the negative reference, the counter is stopped and a count 'T_2' is noted. The times taken to count the duration of the up and down ramps are now in proportion to the respective voltages V_{in} and V_{ref}. By simple geometry it is apparent that:

$$V_{in} = \frac{T_1}{T_2} . V_{ref}$$

Any noticeable drift in the amplifier's offset voltages or in the resistor or capacitor values due to temperature is very unlikely to occur during a conversion. It does not matter at what rate the ramps change in absolute terms, consequently the ratio of the two times is extremely accurate. The voltage reference must, however, be very accurate as it does determine the resolution and stability of the converter. By concentrating design effort on this, and other more minor points, converters have become available commercially with 20-bit accuracy. This technique is also cheap, if slow, and so forms the heart of almost every digital multimeter (DVM or DMM).

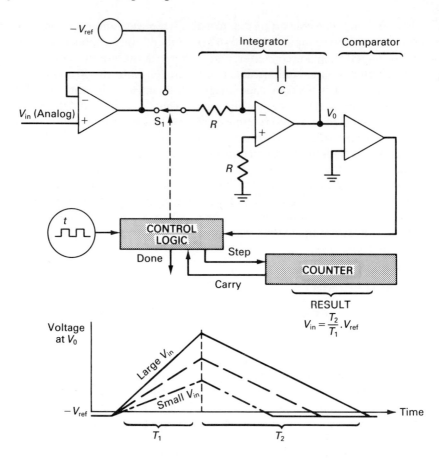

Fig. 6.6 Dual-slope analog to digital converter

The time period for the integration process (T_1) is usually chosen to give specific frequency noise rejection, commonly against 50 Hz mains pickup. One cycle of AC mains takes 20 milliseconds so any multiple of this for the unknown input ramp will minimize noise pickup by averaging over exactly one cycle. It should also be noted that higher frequency noise should also be averaged out by the integrator.

6.4.2 VCO converters

Instead of using time as the intermediate variable, its inverse (frequency) could be used. A voltage-controlled oscillator (VCO) can be used which converts an input voltage to an output frequency which is then measured by a microprocessor. It is quite possible to use changes in component values to alter the frequency of oscillation of the VCO instead of requiring an input voltage to convert. Figure 6.7 shows a practical circuit which includes the transducer as part of the oscillator and thus gives a frequency proportional to the variable. The depth of fluid in the trough alters the capacitance between the plates by changing the dielectric from fluid to vapour and vice versa. The change in capacitance alters the frequency of oscillation of the type 555 astable, giving f inversely proportional to depth. This approach is very suitable for low-cost, relatively low-frequency systems. As another

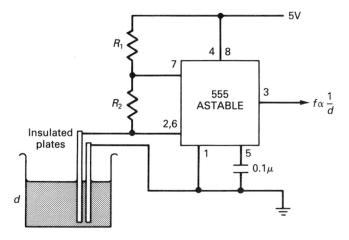

Fig. 6.7 VCO analog to digital converter

alternative, a circuit which produces a repetitive pulse width, proportional to the variable to be measured, could be used.

6.5 Microprocessor-based Converters

There are two aspects to basing converters on microprocessors. The first is that if the converters are externally interfaced circuits then they should be designed to connect to a micro with the minimum of trouble. The second aspect is that it is possible to use a microprocessor to perform all the control aspects of the converter which would otherwise have employed external hardware.

Figure 6.8 shows both an A-D and a D-A converter configured for direct connection to the bus wires of a typical 8-bit microprocessor. For the digital to analog converter (a) the connection is very simple for it is only an 8-bit unit and so matches the data bus. Only vestigial address and control busses are required. A single chip-select ($\overline{CS}$) will suffice, connected, as with store chips, to external address-decoding/selection circuits. The read/write line R/$\overline{w}$ and a clock signal θ_2 ensure the data is latched into the D-A at the correct time, whenever a valid write is made to the chip. After the propagation delay of the D-A the new analog value appears at the output. If the D-A has a longer word length than the data bus can supply, say twelve bits, then two writes will be required, with an extra address line (A_o) to select which half is loaded. The latches will have to be double buffered to ensure that there is a smooth change to the new value and no chance of going via an incorrect, half-loaded value.

Connection to the A-D converter (b) is a little more complex. As the conversion must be started off some extra decoding is needed. For example, writing to the chip might commence a conversion. Completion can be signalled by a bit in one of the data registers. For convenience this can also signal an interrupt on a separate line (IRQ). This in turn could be cleared by reading the new digital value in read operations on the two halves (to give 12 bits plus the *done* signal). The gates which connect the registers to the data bus use tristate logic to permit a bussed connection. This concept is discussed more fully in Chapter 7.

The second way of using microprocessor-based conversion is to have the microprocessor perform all the control aspects of the converter which would otherwise have used external hardware. For applications where relatively low frequencies are to be converted only a digital to analog converter is needed with the microprocessor doing everything else. For slightly higher frequencies the aperture time has to be minimized by including a sample/hold circuit. Figure 6.9 shows flow diagrams for a very simple counter-type converter and for the successive approximation algorithm which is most likely to be used. The hardware external to the micro is shown in Fig. 6.10 assuming the D/A is not available with the micro-adapted interface discussed previously. It is thus interfaced via a Parallel Interface Adapter (PIA) chip which is described in detail in Chapter 12.

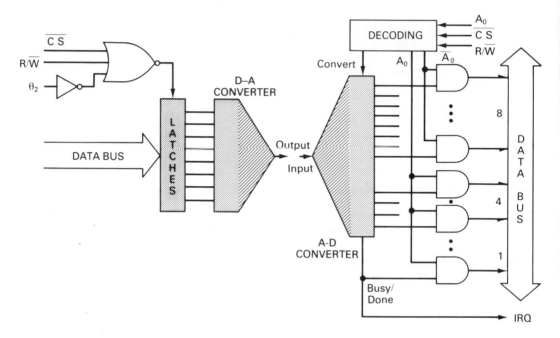

Fig. 6.8 Microprocessor interfaced D-A and A-D

Once the unknown input has been held the successive test values are output to the D-A by the micro, which reads the comparator input to determine what the next test value should be. A program to carry this out is shown in Fig. 6.11.

To cut further the amount of hardware external to the microprocessor, the successive approximation algorithm and its necessary D-A can be replaced by the slower but more accurate dual slope algorithm. Only two output bits and one input bit are required, but for **very** low frequency input signals the sample/hold can be omitted and only one input and one output used. The algorithm discussed in Section 6.4.1 is easily applied to the minimal hardware of Fig. 6.12.

Using the microprocessor to carry out the control functions of A-D conversion obviously reduces the cost, simplifies construction and increases flexibility but these converters cannot possibly operate at the speed of explicit hardware. The successive approximation algorithm takes seventy instructions for an 8-bit (0.2%) conversion. This will typically take 0.25 milliseconds and so the sample time is limited.

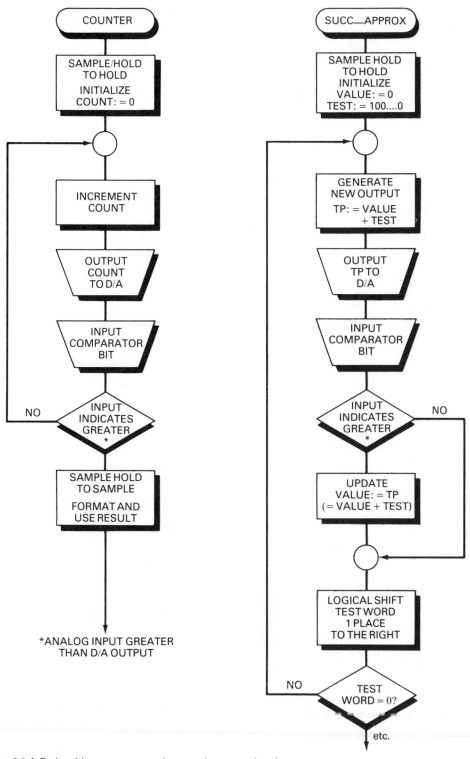

Fig. 6.9 A-D algorithms, counter and successive approximation

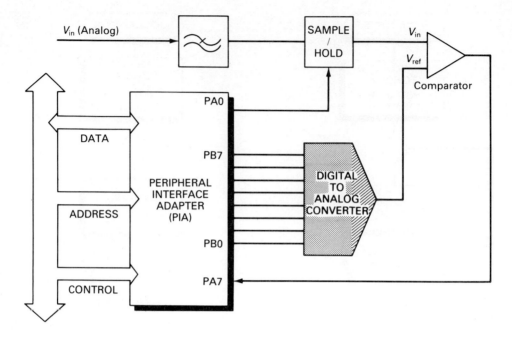

Fig. 6.10 Hardware for programmed A-D converter

```
Test_pattern         DATA     0                  {temporary store}

Succ_Approx          ENTRY
{successive approximation A-D, result returned in register B.}

                     CLR      PIA_data           {S/H to hold, bit PA0 of PIA}
                     LDA      #080H              {pattern for 0.5(Vref) in hex}
                     STA      Test_pattern
                     CLRB                        {zero result so far!}

Succ_Loop            TBA                         {to compute new test value}
                     ORA      Test_pattern
                     STA      PIA_D/A            {output test value}
                     LDA      PIA_data           {input comparator o/p, top bit}
                     BPZ      Compare_low
                     ORB      Test_pattern       {go larger, so add in this bit}
Compare_low          ROR      Test_pattern       {divide by 2, for binary chop}
                     BCC      Succ_loop          {repeat, till test bit off the end}

                     LDA      #001H
                     STA      PIA_data           {set S/H back to sampling}
                     RTS
```

Fig. 6.11 Program for successive approximation A-D converter

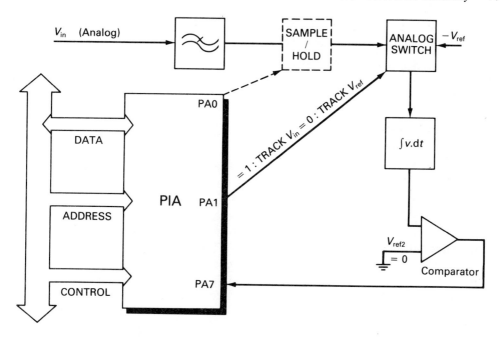

Fig. 6.12 Hardware for dual-slope A-D (minimum)

6.6 Converter Accuracy

The primary error in an analog to digital conversion is the quantization error. Unlike a D-A when a given digital value is converted, hopefully, to its exact analog equivalent, an A-D must take an unknown input voltage and **approximate** to it the nearest of a finite set of digital values. The quantization error is inherent in all digitization and the maximum quantization error is the smallest increment $\pm\delta V$ of input voltage V to which the digital output can be approximated: $\pm\delta V = V/2r^n$ where r is the radix and n is the number of digits. For example, if $r = 2$ (binary) and $n = 8$ then the quantization error is $\pm 1/512$, i.e., $\pm 0.2\%$.

This error can be thought of as noise and from an ideal converter is a triangular wave with maximum (peak) of $+V$ and minimum $-V$. As with most noise sources the average value is zero, but the root mean square value of the triangular wave is $2\delta V/\sqrt{12}$. Thus conversion may be viewed as a signal processing function which adds noise to the original signal by virtue of the quantization process. As this is inherent in the conversion process it could only be made zero by an infinite resolution converter. All we can do is to reduce it to a level commensurate with the overall accuracy required from the system.

The signal-to-noise ratio is easily computed but varies with the signal size in this case. In decibels it is:

$$\frac{S}{N} = 10 \log \left| \frac{2^n/2\delta V}{(2\delta V/\sqrt{12})} \right|^2 = 20 \log(2^n) + 20 \log \sqrt{12}$$

$= 6.02n + 10.8$ for maximum signal or 10.8 for minimum signal.

This is the argument for using a non-linear conversion if a uniform error is required across the range. For this purpose the binary weight D-A used in a successive approximation converter would be replaced with, for example, a logarithmically weighted one.

The dynamic range of a converter is often quoted, and is the ratio of full range to resolution. From the above it can be seen to be

Dynamic range $= 6.02n$ where there are n bits in the A-D

The same errors described in detail in Section 5.6 are carried over with the D-A into the A-D.

- **Monotonicity**
- **Linearity and Differential linearity**
- **Offset and gain errors**
- **Temperature coefficients.**

Of these only the linearity is really affected by the type of converter rather than its detail of design. The feedback techniques based on D-As exhibit worst differential linearity errors at the major transitions, particularly at one half of full range when all bits are required to change state (01111111 to 10000000 or vice versa). The next worst cases are the quarter and three quarter full scale points and so on. The dual slope technique, on the other hand, exhibits very small differential linearity error but shows ordinary linearity errors due to the integrator circuit failing to operate ideally. The differential linearity error is determined by errors in the timing of clock pulses but, as we have noted, counting is very accurate in microprocessors or digital hardware.

Thus to choose an A-D conversion technique, first consider the **speed** required, then the **resolution** and **dynamic range** needed, then the **accuracy**, and finally if there is any **multiplexing** requirement. For speed, flash converters are fastest, followed by tracking, successive approximation and dual slope. For resolution and dynamic range the dual slope converter comes out on top followed by the feedback types and finally flash converters. The same order applies to overall accuracy. Finally, the tracking types are only suitable for non-multiplexed inputs whereas the others are suitable for multiplexed or non multiplexed inputs. Of course, cost will often be the ultimate decider.

7

Interfacing to a Computer

With all the signals in digital form it remains to connect them to the computer and have them read in. The connection point is what is traditionally regarded as an **interface**; however a wider view must now be taken. Definitions of an interface are difficult as the term covers so much. An interface is a boundary between a control device and a connected device or devices which may or may not include controlling logic, such as a transducer, a peripheral or another processor. An interface is the definition of the *logical*, *electrical* and *physical* properties of the boundary, but the definition has to be extended further to **protocols** as any interface these days is a combination of both hardware and software. The balance of hardware and software can usually be varied, one way giving greater speed of operation and the other way reducing the cost of the connection.

The interface does not have to be a single boundary as the definition can be of the visible boundaries surrounding some logic private to, and included in, a so-called **thick** interface. There are not yet international standards for interfaces of this kind.

The organization of the interfaces and the input/output system in general has a bearing on the architecture of the overall system. The use of interfaces, and standard interfaces in particular, also impacts on high-level languages which until very recently have been notable for their inadequate handling of input/output. The use of rigorously developed interface standards could be mirrored by the development of standard language constructs to permit the software parts to be high level rather than the assembly language additions needed for full BASIC, FORTRAN or Pascal input/output.

It is instructive to categorize interfaces by their data handling capacity and their ability to handle single or multiple devices. Such a categorization is shown in Fig. 7.1 and the application areas of each type are apparent. Transducer and converter connection could fall into any category but the more common arrangements will be shown later. Each type of interface in Fig. 7.1 matches a given balance of distance, speed and cost. The shortest interface connects the printed circuit boards in a rack. In such a **backplane** interface it is quite practical to carry all the signals needed on separate wires. At the other extreme, if connecting a device in London to one in Edinburgh, the cost of separate wires for each signal would be prohibitive. Thus the logical signals have to be serialized and transmitted over a single connection. The former (parallel) type has the potential for very high-speed operation while the **serial** type is limited by the large transmission delays and reduced bandwidth inherent in the path.

Turning to the overall style and organization of interfaces and the necessary signals they carry, history gives us some pointers. Previous interfaces have been inconvenient for microprocessor applications because they were designed for entirely different configurations. Traditional computers were constructed with processors which were very expensive, with small stores, and with relatively small numbers of lower-cost peripherals. Thus the conventional hierarchic view of a computer in Fig. 7.2 has the central processor 'in charge'

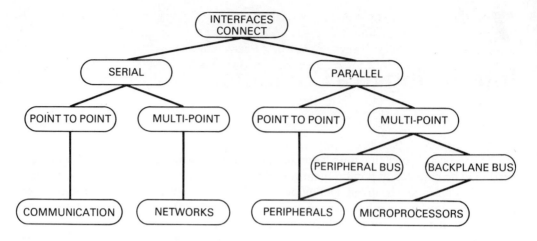

Fig. 7.1 Interface types and applications

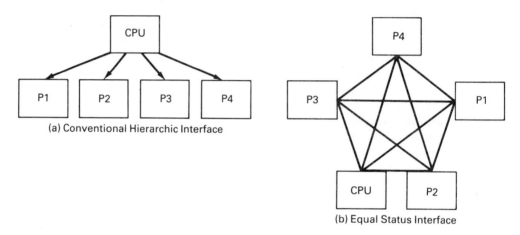

Fig. 7.2. Overall interface organization

with all peripherals (and interfaces) subservient to it. This is reflected in the interface design as the master-slave organization is carried through it to ensure that the processor is kept busy. The same approach was adopted by the common carriers for interfacing to the public switched telephone network and leased lines. The interfaces were/are between terminal equipment (in a slave role) and communication equipment (in a master role).

Modern systems have a different view as the processor is just another part, probably quite cheap, in a total system. Stores are larger, peripherals and their controllers are more powerful and it is quite likely that the one device which can be left idle without concern is the processor. Thus interfaces with a master/slave organization are almost unusable. All the devices have equal status as shown in Fig. 7.2 and can initiate or accept transfers. Of course the separate wiring shown is impractical for larger numbers of devices and so connections may be shared by some form of **Bus** arrangement, either serial or parallel. Thus interfaces must be **symmetric** to allow equal status connections. It is similarly inconceivable nowadays that **any** data pattern should be prevented from being transferred by the interface. Such diverse transfers as binary-encoded transducer data and the keyboard/display interaction with a word processor or screen editor must be accommodated by an interface. Thus the interface must be **transparent** to data.

7.1 Serial Interfacing

Serial interfaces ought to be simpler as they have fewer wires! Unfortunately this is not always the case as many functions have to be carried out over the reduced path(s). Figure 7.3 shows the three main types of serial link. They are different mainly in the amount of control the user has over the **path** linking the serial interfaces of the end devices. On a single site users may lay private wires and so **could** use any interface of their own choice. If the link is longer than the basic interface drivers can handle, then some form of modulation or long line driver (LLD) will be needed and the interface will be restricted to those provided by any available (or new) driver designs. If, however, public roads have to be crossed then, as in the United Kingdom, there are probably monopoly-holding organizations, often known as Common Carriers, involved. Any link must be provided by them and they will insist on a particular interface to either a leased line, to the Public Switched Telephone Network (PSTN), or an Integrated Services Digital Network (ISDN). The user is given no choice. As there is no choice in this case then, to widen their markets, manufacturers adopt this interface and, even if it is not really satisfactory, use it for all the other cases.

The simplest of serial links is called a **Simplex** or **Channel** connection. It provides a single path in one direction only and involves a driver at one end (T_x) and a receiver circuit at the other (R_x). Any electrical signal is conveyed by the movement of electrons and (unless we could find something to produce an infinite source of electrons) a return wire is required to complete the path. If the return wire is grounded and the information is sent by putting an absolute voltage on the signal wire (+10V or −10V for instance) then the transmission is said to be **unbalanced**. It is not the fastest way of using the two wire channel. If the voltage put on the signal wire is only positive (or only negative) the signalling is **Unipolar**. If both

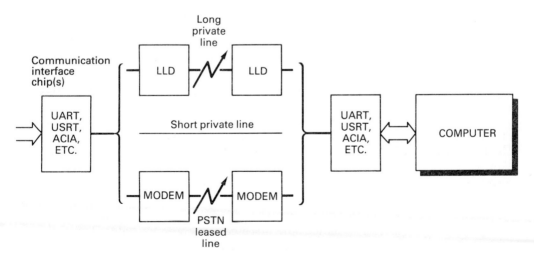

Fig. 7.3. Serial data transfer interfaces

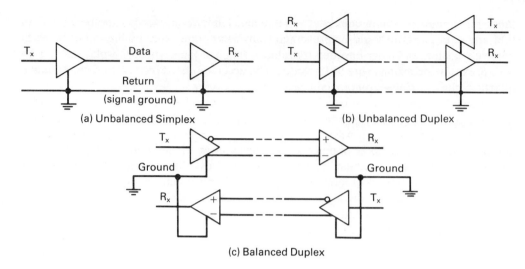

(a) Unbalanced Simplex

(b) Unbalanced Duplex

(c) Balanced Duplex

Fig. 7.4 Serial data transfer connections

positive and negative voltages are used it is **Bipolar** signalling. Figure 7.4a shows an unbalanced simplex connection.

It is useful to communicate in the reverse direction as well and an extra wire, driver and receiver can be added as in Fig. 7.4b, giving an unbalanced **full duplex** connection. A single path could be used for bidirectional communication if operated consecutively rather than concurrently. The driver and receiver pair at each end connect to the end of the wire via a switch. This is called **half duplex** communication but is not now a common choice. The performance of a channel and the speed of communication can be significantly increased by two methods. The first is to terminate the two-wire path correctly to avoid reflections as described in detail in Section 10.2. The second is to use the two wires in a more ingenious way. Instead of earthing the return path and driving only one wire in an unbalanced fashion, signals can be driven on both wires. For example a **true** signal might be +5V on one wire and −5V on the other. The false signal would be the reverse, −5V and +5V respectively. This allows us to generate a 'shock wave' signal which can be detected more rapidly and followed more closely by the next bit. One must observe, though, at each channel now needs two wires as a common return wire cannot be used, but the extra cost is offset by the greater speed. The full balanced duplex connection is shown in Fig. 7.4c and is discussed further in Chapter 10.

Now we can consider the logical signals which we need to transfer across the interface. For a serial point-to-point interface there are five types of signal needed and shown in Fig. 7.5. The requirement for **return** paths has already been discussed as has the obvious need for **data** wires in each direction for symmetry ($T_x D$ and $R_x D$). To make sure the data is correctly read from the data wires some form of synchronization will be needed. This is discussed fully in Section 7.2 but for now we can assume a **clock** wire ($T_x C$ and $R_x C$) paralleling the data wire in each direction to show when the data is valid (the centre of each data bit). When the interface is started up or first used it may be in any state so logical signals are included to allow a **reset** in each direction (T_{RS} and R_{RS}). This will also allow an orderly return to a known state on either side of the interface should a serious error occur. Finally there are a number of functions required to control an interface during and between data transfers. They used to be handled by control codes—taking and using some of the data patterns. This meant that those codes could not be used for data. As we now insist on transparent interfaces another technique must be used and for now we can include a **control** wire in each direction to complete the necessary signals for a serial interface (C and I).

This would seem to imply that in theory we need a **16**-wire interface (8 signals and 8 returns) for the ideal serial interface. This would appear rather a lot for 1 bit of data in each direction and there are many ways to cut the number, using a variety of techniques to merge logical functions onto fewer physical links. For two concurrent high-speed channels the minimum must be **4** wires (2 data plus their returns) if all other functions could be merged onto them.

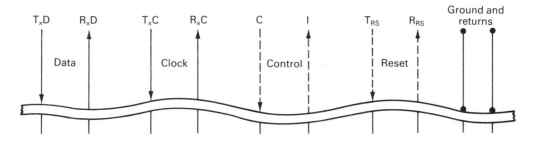

Fig. 7.5 A serial interface

Synchronization techniques are covered in Section 7.2 and two of them provide for single channel (no separate clock channel) data transfer. To merge the control and reset functions requires an additional approach and there are three possibilities beside using a separate control channel:

- **Time division multiplexing**
- **Code insertion multiplexing**
- **Multilevel signalling (amplitude division multiplexing)**

Time division multiplexing can be operated to mix data and control in an imposed interleaving. There are considerable overheads of course as the time slicing required will provide slots for control and data whether they are needed or not. More complex demand-based methods are used for long-distance time sharing but the simpler code insertion mechanisms are better suited for the short haul of an interface connection.

There are two methods of multiplexing control information by code insertion in common use: **bit insertion** and **byte insertion**. For byte insertion one of the byte patterns is chosen as an indicator that control information follows. If this pattern occurs in data then an identical pattern is inserted following it before the next byte. Thus at the receiving end if a single or odd occurrence of the chosen pattern is met then control information follows. If there is an even number of the chosen pattern then only data is involved and every even one is discarded as it was inserted. For bit insertion again a byte pattern is chosen but it has specific characteristics—usually 01111110. As this flag includes six ones in a row then whenever that sequence occurs in data it is prevented by insertion of a zero after five ones. At the receiver after five ones are received the next two bits are examined. If the first is a zero it is discarded because it was inserted, but if it is a one then the next bit is examined. If it is a zero the control flag has been detected but if it is a one then seven ones (effectively a continuous stream of ones) have been found and this is an error signal causing a complete reset. This gives bit insertion an advantage as both control and reset are now multiplexed with the data. Consequently, bit insertion is currently the most popular method.

Multilevel signalling obviously allows for similar mixing of control and data information with different amplitudes allocated to each role. The problem is the added complexity of drivers and receivers and the inherently lower reliability. Any signal transmitted over a distance is attenuated and depending on the channel also suffers burst attenuation (noise) or frequency-dependent attenuation. Some simple systems are operated, however, using a bipolar signal to give zero and one for data and no signal to indicate some other condition e.g. a **break** or reset signal.

7.2 Synchronization and Timing

A variety of techniques exist to ensure that a receiving station reads data at the correct times. A transmitting station must put data onto the data channel(s) and then change it to the next state(s) and so on in a continuous stream of bits (or bytes or words) to form any meaningful transfer. As has already been observed, any data pattern or sequence of patterns may be transferred over an interface. This implies that no data pattern or change in data pattern can be used to indicate when data is valid. Multilevel signalling would allow a change to a higher voltage (from the rest state) to indicate one value, with a change to a lower voltage indicating the other. However, if we restrict ourselves to binary signalling then there are four basic methods to indicate the validity of data or, in other words, to synchronize reception with transmission:

- **Synchronous transmission**   —with an explicit clock
- **Enchronous transmission**   —with an embedded clock
- **Isochronous transmission**   —with two similar clocks
- **Asynchronous transmission**   —synchronizing without clocks.

There is a problem with this terminology—many manufacturers use the term 'asynchronous' to refer to the case using similar clocks at either end (i.e. two clocks!). True asynchronous operation uses no clocks and signals in some other way.

Synchronous transmission uses a clock signal produced by the transmitter and sent along a separate channel in parallel with the data channel(s). Figure 7.6 shows a changing data stream with the clock (option A) being true if the data is valid and false if it may be changing. This is simple but the clock signal has to change twice as fast as the maximum rate of change of data (e.g. 010101 . .). Option B uses a single **edge** to mark a point where data is known to be valid—close to the centre of the bit—and overcomes this problem. This is called **transition** clocking and is a special case of transition signalling. Transition signalling is not common at present. It uses two lines and signals a 'zero' by a change (a transition) on one line and signals a 'one' by a change on the other line. With transition clocking, data and clock can change at the same (maximum) rate and somewhat greater skew between the two can be tolerated.

Thus synchronous transmission is the **fastest** method of transferring data as it uses only one edge (or change) per bit and we can nominate this as 100% efficiency. It does use two channels, however, and thus the total channel efficiency is only 50%.

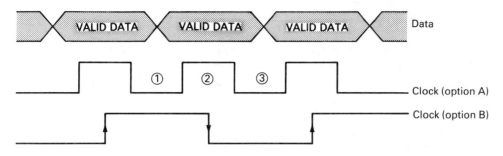

Fig. 7.6 Synchronous data transfer

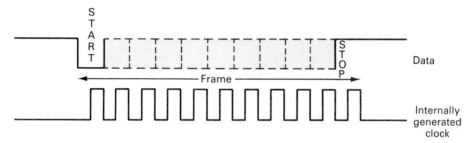

Fig. 7.7 Isochronous data transfer

The same total efficiency can be achieved by simply alternating timing and data information on a single channel. This embeds the clock into the data giving **enchronous** transfer. There are a multitude of different ways of doing this called variously: phase encoding, Modified Frequency Modulation, Manchester encoding and so on. They employ clock edges at regular times between each data time but differ in using the state, change or absence of change, or direction of change to determine whether the data is zero or one. In some it is the data edges which are regular and guaranteed to give a change, and the intermediate change or absence of change serves to get the correct state for the following change indicating data. These schemes vary in complexity of implementation and in theoretical reliability. There is a maximum of two edges per data bit (minimum one) and so the efficiency on the single channel is only 50%. The data is transferred at half the rate of synchronous transmission but only a single channel is needed and there is no skew problem. Each data 'cell' could stretch or shrink by up to 25% and still (just) be recovered. These techniques are common therefore for magnetic tape and disc encoding as well as in local area networks.

Over long distances and where bandwidth is limited, greater efficiency is always sought. Both synchronous and enchronous methods have given up half of their available bandwidth to carrying synchronization. Figure 7.7 shows the **isochronous** approach where the receiver is assumed to have a local clock running within 10% accuracy of that of the transmitter.

With this addition, the amount of synchronization carried in channel can usually be cut down to 20%. Instead of synchronizing to every bit with the explicit synchronous or embedded signals we assume that the receiver will be able to 'look after itself' for say eight bits and only synchronize for two (1 guaranteed edge!).

If the data line rests in the 'one' state, for example, then a frame of data commences with a change to the 'zero' state indicating the beginning of the **start** bit. Following this bit are eight similar width periods to accommodate any pattern of eight data bits. Then there has to be one bit time back in the rest state (one) to ensure that the next initial 1/0 change can occur. This is a **stop** bit. It is obviously easy for the transmitter to create this arrangement using **its** clock to fix the timing. It's a bit more complicated for the receiver though. Remember that it has no direct timing information or validity signals from the transmitter—only a clock running at a similar frequency to that of the transmitter, or close to it. The first thing the receiver sees is the $1 \rightarrow 0$ (one to zero) edge of the start bit. Ideally this would be used to start its clock which would then clock in the data bits. Sadly this is difficult for continuous clocks.

Instead the receiver runs a clock not at the similar frequency to the transmitter's but at a multiple of it: usually four, eight or most commonly sixteen times the transmitter clock. When the leading edge of the start bit arrives it is gated with the clock, signalling on the nearest following clock edge. Counting eight clocks from this signal, the centre (approximately) of the start bit is checked to ensure it is a genuine start bit and not noise. Subsequently every sixteenth clock will hit near the centre of each data bit and even if the clock is 10% faster or slower the eight data bits and stop bit will still be synchronized. The overhead of the two bits (stop and start) to guarantee the 1/0 synchronization between them is quite small and the only drawback for very high speed is the need for the **local** clock to run much faster than the data rate.

The isochronous transfer mechanism described above is the most efficient single channel two-state method for ensuring reliable transfer, having 80% efficiency. It can be thought of as a special case of enchronous encoding, using a particular receiving mechanism. It can also be thought of as being asynchronous between bytes (frames) and synchronous by bit within a frame; perhaps this is why manufacturers sometimes call this asynchronous.

The problem with **all** the preceding techniques is that the receiver is required to accept data at rates determined solely by the transmitter. For a standard a set of convenient speeds may be defined and the ends pre-agree which speed to use. Only if **no** clocks are involved, i.e. if asynchronous transfer is used, is this problem completely solved. This is particularly important in multipoint interfaces where devices of very different capabilities could be involved.

An **asynchronous** point-to-point transfer control needs two channels in addition to the data channel(s) as shown in Fig. 7.8. The channel from the transmitter to the receiver indicates the validity of data on each rising edge in a broadly similar fashion to Fig. 7.6, option A. It is the addition of a reverse channel—from receiver to transmitter—which gives the flexibility. This channel indicates when the receiver **is** ready to accept data and then when it **has** taken the data. The sequence of actions starts with the transmitter *waiting* until the receiver indicates on its *RR* channel that it is ready to receive—as at (1) in Fig. 7.8. The transmitter puts out its new data and then asserts that it is valid on its *DV* channel—as at (2). The receiver may take whatever time it needs to take the data, then indicates it has done so by lowering its *RR* channel signal—as at (3). This also indicates that the receiver is not ready to accept another data item as it may still be handling the data it has just taken. The transmitter then removes its data, asserting that it is no longer valid by lowering the signal on its *DV* channel—as at (4). The whole cycle then repeats at whatever rate the receiver will accept and the transmitter will run at. The final edge from the transmitter is, strictly speaking, unnecessary but it is usual to give the receiver a check on the transmitter handshake. It can be seen that this is another case of the transition signalling mechanism described earlier, but with receiver speed-control. The efficiency in the best case is thus

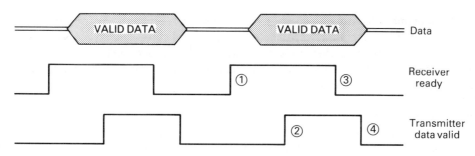

Fig. 7.8 Point-to-point asynchronous transfer control

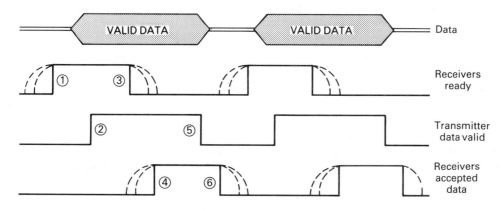

Fig. 7.9 Multipoint asynchronous transfer control

determined by needing three edges per transfer plus any transmission delay between them. This is also spread over at least three channels (one for data) giving data channel efficiency of 33.3% and total channel efficiency only marginally over 10%. It is, of course, much better suited to parallel transfers and does give the great advantage of an *automatic transfer rate* achieved, as described, by *receiver* control.

If transfers are needed on a **multipoint** connection then the first three mechanisms need no alteration. As at any instant only one device transmits—all the others being receivers or uninterested in the transfer. An explicit clock or embedded clock is then quite satisfactory. Because the source controls the transfer rate the receivers just have to keep up.

For an **asynchronous multipoint** mechanism, because it is the receiver which controls the transfer rate we must then ask 'which receiver will control the transfer?'. There are two signals which must come from these receivers. The *last* device to become ready to receive must tell the transmitting device that all receivers are ready. The *last* device to accept the data must tell the transmitting device that all receivers have taken the data. These two signals *might not* come from the *same device* and try as we may no two-wire arrangement of the form of the point-to-point asynchronous control can be made to work.

Three signals, shown in Fig. 7.9, are required for the general multipoint asynchronous transfer mechanism: **receivers ready for data**, **transmitter data valid** and **receiver data accepted**.

The transfer mechanism proceeds as follows. First, all receivers indicate their readiness to accept data by releasing their switch on the receiver ready line. This is shown by the dotted lines in Fig. 7.9. The last device releasing the line causes it to become *high* as shown by the solid trace (1). The transmitter then puts the data onto the data lines and sets the data valid signal *high* (2). The receivers then each independently lower their ready signal (3). Although this could be done at other times it is convenient to split the transactions into the two self-contained handshakes. The receiving devices then indicate that they have accepted the data (4). Again the last receiver to release the line causes the 'accepted' signal to become *high*. The transmitter sees this and lowers the data valid signal (5) prior to removing the data. Finally the receivers return their accepted signal to *low* (6) and the whole sequence repeats. There are alternate sequences of similar actions which would achieve the same end.

By transition encoding from the transmitter one can arrange to need only five edges to execute a transfer, although three wires are always needed in addition to the data lines. Thus the data channel efficiency is 30% and the total channel efficiency only 5% for one data channel. All the same advantages apply as for the point-to-point case.

7.3 Parallel Interfacing

There are strong arguments for serial interfacing over long distances or if network switching is involved. The number of interconnection pins is fewer and so the cost is minimized. If the interface lines are short and no switching is involved then strong arguments may be advanced for using parallel data paths. If we transfer data at a given rate on one wire serially organized then using eight parallel data paths will give an eight-fold increase in transfer rate.

The main difference between serial and parallel interfaces, apart from the number of data lines, is that if we can afford them then we can afford separate lines for synchronization, control, etc. So for a short point-to-point parallel interface we may have many pins carrying all the signals we need separately. For example a common printer interface has 8 parallel data lines yet uses a 36-pin connector.

Although there is a place for point-to-point parallel interfaces for local printers, etc.,

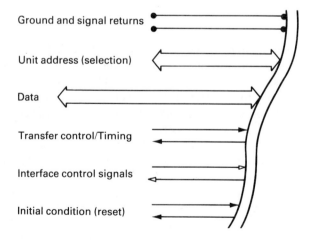

Fig. 7.10 Multipoint interface signal types

more interest is focused on parallel multipoint interfaces. The necessary signal types are shown in Fig. 7.10, the most noticeable being the **unit-address** lines. These permit selection of the chosen device and are the main logical difference from point-to-point interfaces. Physically, drivers will have to cope with a multiple load and connectors with some form of **daisy chain**.

When the interface is shortened so that it fits inside a single case or rack, for example to make a computer from various printed circuit boards, then it is usually called a bussed interface. A backplane card or backplane wiring carries the signals, hence the other common name: backplane bus interface. Much of what follows about the shorter interfaces also applies to the longer parallel interfaces of this section.

7.4 Bussed Interfaces and Buffering

Two serious problems have just been quietly glossed over: **drivers** and **connectors**.

The line drivers of a bussed interface will have to be able to drive multiple loads. They have to drive the inputs of the receivers of other devices, but the drivers of the other devices also present some load. There is also the case when numbers of drivers drive at the same time, for example the 'ready' and 'accepted' lines discussed before. In this case it is not obvious to which logic level the outputs will end up being driven. Fig. 7.11a shows the problem for two LS (low-power Schottky) or standard TTL drivers connected to the same line but driving opposite signals. Current will flow from the power supply through the two 'on' transistors (one in each chip) to earth. As there is nothing to limit the current it will be excessive and serious damage may occur. The same problem occurs with CMOS logic but the currents are more than an order of magnitude less and so actual damage is not likely. The ordinary outputs of NMOS logic have a resistive pull-up (i.e. a passive pull-up) and a transistor to drive the output low (i.e. an active pull-down). This means that they can be connected in parallel as no two active circuits can oppose one another. Of course, for ordinary logic only one output drives a number of inputs. It is only if we need to 'bus' a number of possible outputs that we will need to do better.

The first standard solution is to copy the NMOS type of output and dispense with the 'active' pull-up transistor. Special driver outputs are provided with just a transistor with its collector unconnected (or open). The 'bussed' connection is made by connecting all the 'open collectors' of the drivers together and to a single resistor pull-up to the power supply. This arrangement is shown in Fig. 7.11b. It is simple and provides the necessary AND function for multipoint asynchronous transfer control. The inconvenience comes in the external pull-up. In which circuit should it be located, or should it be an extra component? This arrangement also causes the signal rise time to be slower than the fall time, because there is an active pull-down, but only a passive pull-up. It would thus be desirable to have both active pull-up and pull-down transistors in the output of the drivers yet have no other transistors operating from other drivers on the bus. Such an arrangement is possible if we include some extra logic.

The full circuit of a driver is shown in Fig. 7.11c. It has a normal logic input and an extra enable/disable input. If the enable is true then either one or other of the output drivers (pad drivers) will be turned on by the logic input. This gives a '0' or '1' output. If, however, the enable signal is false (disable) then neither output driver can be turned on and the output simply floats. It can then be driven easily by another driver enabled on the bus. This circuit is called a 'tristate' driver because it has three output states: driven zero, driven one, or a high impedance undriven state. Tristate outputs are easily fabricated in CMOS logic and are

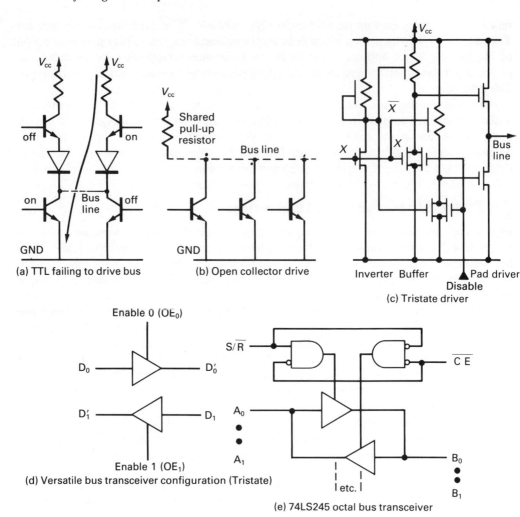

(a) TTL failing to drive bus   (b) Open collector drive

Inverter Buffer   Pad driver
Disable
(c) Tristate driver

Enable 0 (OE₀)

(d) Versatile bus transceiver configuration (Tristate)

(e) 74LS245 octal bus transceiver

Fig. 7.11 Bus line buffering and driving

particularly suited to address and data drive buffering. This is because they are operated deterministically (we always know which device is to drive the address or data bus and in which direction). Various configurations of this type of buffer are made in integrated circuit form either as transceivers or as isolated (versatile) pairs of drivers as shown in Figs 7.11d and e. One must not confuse a tristate driver with three-level or ternary logic circuits which can drive one of three real states. Most bus drivers in current use are *not* capable of driving the low impedance loads of the bus with correct matching and so rely on reflections to build up the correct voltage over a short, but finite, period of time. This problem is discussed further in Sections 8.4 and 10.2.

On the connector side a printed circuit card can either plug directly into a socket on the backplane (or 'mother card') or itself carry a connector (plug) which fits into a connector (socket) on the backplane. This is an 'indirect' connection system. The use of direct gold-plated fingers on the sides of a printed circuit card has problems. They are not as reliable, cannot be as densely packed, and cannot carry the current of an indirect connector. They are also more prone to misalignment. With badly designed cards they can even be plugged in the wrong way around. Indirect connectors, particularly the DIN standard 32 pin/row: 1/2/3 row connectors, are definitely the preferred choice for backplane connection. They are polarized to ensure the correct alignment and have good electrical characteristics. This is discussed further in Chapter 10.

We can now see what signals are required on the backplane bus over and above those of a parallel interface. We still need grounds, signal returns and data lines. It is also likely that a backplane will not only handle input/output devices but also store and perhaps even co-processors. For these extra functions more data lines (to the processors' word width) and address lines to the full real address space of the processor, are to be expected. It will be convenient to distribute power supplies along wider (higher current) paths on the backplane. This is not normally the case for ordinary parallel interfaces. The basic mechanisms for timing and transfer control will remain the same but there are usually some extensions. For example, a continuous clock for timing, other than for transfer control, e.g. for real-time timing is often added. The initial condition handling is usually made more extensive to cope with sequencing for power-on and power-off over the whole backplane. A general backplane bus is shown in Fig. 7.12.

The most major extension is, however, in interface control. As most backplanes are required to operate at higher speeds (i.e. higher transfer rates) and with multiple masters (controllers) the mechanism for choosing the next controller must be efficient.

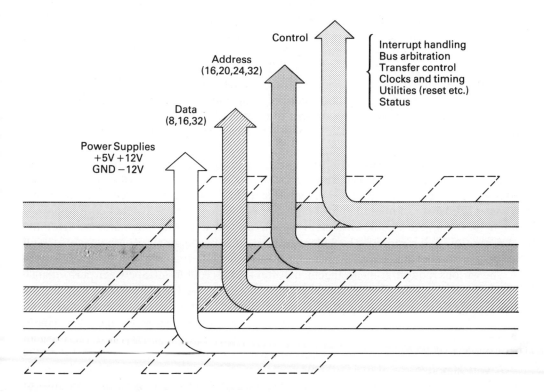

Fig. 7.12 A backplane bus interface

There is a spectrum of possible solutions ranging from random choice to explicit, or **deterministic**, transfer of control.

- **Random transfer of control (e.g. CSMA/CD)**
- **Request/Grant with variable priority**
- **Request/Grant with fixed priority**
- **Explicit transfer of control**
- **Deterministic cycle of control**

Random transfer is of only theoretical interest for backplane use. It is used in multipoint serial connections like the Ethernet local area network. The basic operation is to monitor the bus for no activity and then to transmit. This is carrier-sense (monitoring) multiple access (i.e. CSMA). However, because of the finite speed of transmission (close to the speed of light) and the distance between stations, there is a quite finite chance that two stations seeing the bus empty will transmit. This will corrupt their data and is called a collision. It is detected (i.e./CD) by each station monitoring the data it itself is sending. Following the collision the colliding stations put a jamming signal onto the bus so that every station knows to ignore the data, and then backs off for a random period. The random period is based on the station's address, the frequency of collisions and other factors designed to minimize the likelihood of repeated collisions.

This system has a number of drawbacks. It performs worse as the load increases until it 'stalls' completely. There is a minimum transfer size in number of bits to ensure that if there is a collision it is seen as a collision by **all** stations. This minimum increases as the bus length increases and for Ethernet is about 500 bits in series. It is very wasteful of bandwidth and cannot give a guaranteed real-time response to any controller needing the bus. So CSMA/CD could be used for local area networks, even though it is not exceptional for them, but is quite inappropriate for high performance backplane arbitration.

Explicit transfer of control is much better if the cycle of needs for control is known or can easily be determined in some other way. The IEEE488 interface discussed in Chapter 8 uses this system. It has the great advantage of simplicity. The current controlling device addresses the next controlling device and explicitly instructs it to take over. The disadvantage is for the device that is never offered a chance or worse if a failure means no other device is ever given control

A deterministic cycle passes control from device to device giving each a turn to use the bus or pass it on to the next. This technique is also used in serial buses when it is called 'token passing'. Because it is deterministic, merely setting a limit to the maximum deviation of a transfer guarantees a minimum for real-time performance. The maximum performance is not achieved unless all devices actually want to execute a transfer when they are given the opportunity. This is rare for a backplane bus as a single processor may continuously access a single store area, with only minimal I/O, for a long period.

The various request/grant arrangements offer the best compromise between performance and fairness. In its simplest organization three signals are used: Bus **Busy**, Bus **Request** and Bus **Grant**. All devices connect to the Busy and Request signals. The grant signal starts from one end of the bus and is opened to pass through a logic gate at each device like a daisy chain. A device wishing to use the bus observes the busy signal false and puts up its request. Again two or more devices may request at the same time.

A very simple piece of logic sees the request-true, busy-false, and sets grant true to the first device in the daisy chain. If this device requested then it takes the grant and continues to pass a false grant on to all the other devices. If it did not request then it sets its output grant true to the next device and so on. The first device in the chain to have requested and receive the grant holds it as described and sets bus busy true. This whole mechanism is very fast—only a few logic gates delay—and foolproof. After a transfer is complete it drops bus busy to clear the grant and restart the cycle.

The difficulty is that the priority of service is fixed by the physical position on the backplane. It is also possible to lock out the lowest priority device(s) continuously. For a single processor this can be avoided by making it the lowest priority device. For multiple processors sharing a bus the problem remains.

The inclusion of more complex arbitration logic and priority levels based not on position but choice can be arranged to improve the system. This requires extra priority signals on the bus and provision should be made for this in any backplane to support multiple processors.

The final signals to include on a backplane bus are service requests or interrupts and their associated vector address lines. The whole operation of interrupts and interrupt structures is covered in Chapter 11.

If a bussed backplane is to have any future it must already address and solve all these points. It must be done in a flexible manner so as to accommodate various host processors, store sizes and widths, speed of operation etc. and permit its widespread adoption.

An attempt to recap on the need for interfaces at various levels and the requirement for standardization is shown in Fig. 7.13. Inside a very large-scale integrated circuit (VLSI) there are buses and interfaces. Individual manufacturers support standards in their companies for bus arrangements cell layouts, clock phases, precharging levels etc. It seems

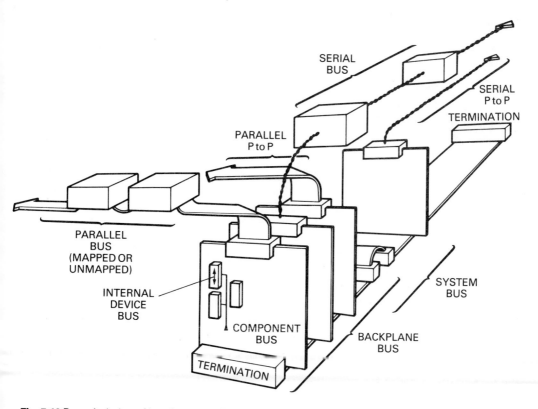

Fig. 7.13 Recapitulation of interface types for standardization

unlikely such standards will become or need to become international in the near future. One level out, on a printed circuit card, is the first place where standards already exist. JEDEC standards exist for chip pin placement for store chips and others. The inter-integrated-circuit bus (I^2C) bus has only two lines. Serial data and serial clock are used to give inter-chip transfers between up to 128 devices at 100 kbps data rates. For point-to-point communication between transputers, isochronous data rates of 10 Mbps are achieved. There is further discussion of these mechanisms in Chapter 13. Much work remains to be done at this level so that standard empty cards can take a wide range of processor, store, and I/O chips. Such a component bus would be useful indeed. The backplane bus has been the subject of more 'standardizations' than any other interface area. Most are inadequate at best. A small, well-designed range is all we need, not dozens of different bus standards as we have at present. The more extended 'system bus' simply permits the extended electrical specifications to allow the spanning of a number of racks. The parallel bus is common for connecting multiple peripherals at present. They may be replaced by serial busses in future as much effort is being put into these for high-speed local area network use. The parallel point-to-point and serial point-to-point interfaces will always have a place when their simplicity is desirable. The serial point-to-point interface is the only possible candidate for a truly universal interface.

It is as well to remember that an interface is not just a plug and socket. As is seen in data communication layers of protocol are often needed to ensure compatibility between two manufacturers' devices. A seven layer model has been fixed with definitions of the lower levels agreed, yet the bottom-most one is still not satisfactory.

8

Standard Interfaces

It is surprising to find that one of the most common physical interfaces in process control applications has no agreed international standard. The 20 milliamp (mA) current loop has many advantages, not least its simplicity, over other interfaces for lower speed data transfer. With this system, information is transmitted on two wires for simplex or half duplex operation, four wires being needed for full duplex. Connections are often made by screw blocks. The information is transmitted by a current with 4 mA representing a zero and 20 mA representing a one. The small current flow for a zero distinguishes it from the break or reset condition of no current which conveys only the information that information is not available. The source of the current is commonly a DC supply of from +5 volts to +32 volts. The use of current as the information carrier removes problems due to voltage drops in the line and it is resilient against induced voltage noise. The drivers are designed to maintain the correct current constantly. An example of this occurs in the temperature sensors used in Section 15.3. As will be seen loads may be placed in series to give a composite input or to drive a number of outputs. Of course, the number can only be increased up to the voltage limit of the driver. The noise rejection is also improved as contact resistance and thermo-couple effects in the path do not affect the signal. The final benefit of the current loop is that there is always a small current flowing (4 mA in the zero state) which can be used to supply power remotely. It does not matter which end of the loop supplies the current, of course, so the receiver can supply current both to transmit information **and** to power the devices of the source. With the reduction in device current consumption of complementary metal oxide silicon (CMOS) and liquid crystal display (LCD) technology, this is quite practical. There is also a choice if either end should be allowed to 'float' with respect to ground. In fact both ends could float but this is expensive to arrange and is very rarely needed. As the loop may have more than one receiver it would seem sensible to have the receivers floating (optically isolated, if needed) and a defined side of the transmit station grounded.

Even though the maximum data rate of the current loop is not high, it is a very convenient interface, yet there is no international standard. Why not?

Well, reread from the start of this chapter and consider the problems of standardization. For a standard to be any good, it must have the following features:

- **Completeness** —of electrical, physical and logical definition.
- **Flexibility** —connect a wide range of devices.
- **Simplicity** —of definition, structure and use.
- **Symmetry** —no specialized units or roles.
- **Transparency** —any data pattern.
- **Security** of transfer mechanism.

It also requires definitions of the selection or allocation of sub-roles to the devices and of the protocols for higher levels e.g. the acceptance or rejection of requests. It must have a reasonable implementation cost having regard to its use in other equipment.

Finally, the standard must have international acceptance by both the international user community (you and me) and by one of the international standards organizations. The latter guarantees the standard cannot be 'got at' by any particular manufacturer for commercial reasons. Two terms are used to express acceptance by a manufacturer that their device adheres to a standard. A device is **compatible** with a standard if it adopts the signal properties, connector characteristics and protocols at the specified interface to permit information interchange. A device **complies** with a standard if it completely incorporates all aspects of the standard in the equipment specification, and meets all the restrictions of the standard. The difference is that compatibility only seeks to use the standard whereas complying with a standard means adopting it in spirit and in letter. One further term is used by the standards-making bodies. When they judge that a particular product complies with a standard they say that it **conforms**. This has given rise to a whole field of conformance testing to determine when standards have been met.

For an example, if no definition appears for a particular feature then **no** assumption may be made and all possibilities must be handled. Thus manufacturers who impose more restrictive protocols on a standard do not comply with it. An example of claimed 'compatibility' are all the terminals which have appeared in the past few years with female V24/RS232C connectors, but having data transmitted OUT of pin two. It's like buying a toaster with a socket on the end of the lead!

There are actual faults in many standards, and not just ones left for historical reasons. Most are attributable to one of the following deficiencies:

- **Lack of specification or over-restrictive specification.**
- **Complexity of definition or use.**
- **Permission for alternatives in arrangement or use.**
- **Lack of acceptance by users or international standards organizations.**

Some examples of these are well demonstrated by the IEEE 696 (S100), IEEE 583 (Camac), and BS4421 parallel interface standards.

The 'international standards organizations' are quite diverse. Under the auspices of the United Nations Organization comes the International Telegraphic Union (ITU) and under it the Committée Consultatif International Telephonique et Telegraphique (CCITT). The autonomous International Standards Organization (ISO) is a gathering of the separate national bodies such as the American National Standards Institute (ANSI), the British Standards Institute (BSI) etc. ISO promulgates agreed versions of their standards. In the European Economic Community (EEC) there is the European Computer Manufacturers' Association (ECMA) and in the United States there is the Electronic Industries Association of America (EIA). Under the auspices of the International Electrotechnical Commission comes the committee for European normalization (CENELEC). Finally, during the mid 1970s the Institution of Electrical and Electronic Engineers (IEEE), the American equivalent professional body to the British Institution of Electrical Engineers (IEE), observed the lack of standardization in interfacing, and started promoting their own standards. That there is a vast array of standards is borne out by the first work in the bibliography for this chapter. A compilation of data communication standards in 1982 including no parallel or bussed interfaces or any of their protocols runs to over 1900 pages! It should also be remembered that an interface is more than just a plug and socket, standards being needed for all the higher levels of protocol as well.

8.1 Serial Point-to-Point Interface Standards

The serial interfaces carry only a single bit (in each direction) at a time and so one would suppose that only a few pins would be needed on a connector. The most widespread one has twenty five pins! The V24 standard was defined in the days when copper wire and connector pins were cheap and switching logic expensive. It is still found on every boxed micro-computer though the logic has become vanishingly cheap. A complete microcomputer chip can now be had for less than one of the connectors. So any problems are historical and we must live with them until a better standard is found. The X21 interface has no such excuse though, as it was defined to serve modern high-speed data links.

8.1.1 CCITT V24 (EIA RS232C) standard

This standard was first defined in the late 1950s although it has been revised since. Its basic premise was to separate out the necessary data, control and other signals and place each on a separate pin of the 25-pin 'D'-type connector. This uses lots of pins, drivers and receivers but needs no switching logic.

The signals defined for the pins, shown in Fig. 8.1, are split into **required** signals (which must be present) and **additional defined** circuits which are there to support a range of additional services. One interesting feature of the standard is that it is only defined to connect Data Terminal Equipment (DTE) to Data Communication Equipment (DCE). No attempt was made to achieve symmetry as the DCE was always supplied by the monopoly-holding PTTs or common carriers. It is a 'them and us' standard.

The connector itself (one of a range of 'D' connectors with different numbers of pins) is quite reasonable. It can be board- or cable-mounted and uses soldered or crimped pins. It is reasonably strong, so the pins don't fall off that often.

The first problem arises with the electrical specifications. The voltage ranges for trans-mission and reception are shown in Fig. 8.2. These are inconvenient for two reasons. Microprocessors and their associated chips are currently powered from single +5V supplies. The V24 voltages are higher and also bipolar so extra supplies will have to be generated. The receiver-end load impedance is even more restrictive. It is defined to be between three and seven kohms which is far higher than the characteristic impedance of any cable. The transmitter output impedance is defined to be more than 300 ohms—again far too high. Transmission is also unbalanced. Together, these limit the maximum data rate to only a few tens of thousands of bits per second, as is explained in Chapter 10.

Turning to the arrangement and use, the interface caters for two types of synchronization. In the required-circuits mode transmission is handled isochronously. If synchronous trans-mission is required then pins 17 and 15 are used to carry the receive and transmit clock signals. They both originate from the DCE and this makes differential detection of phase-shift keyed signals in a modulator–demodulator (modem) easier but loses symmetry. Sometimes the transmission speed can be supplied from outside a modem (an input to the DCE on pin 24). This is only used for onward linking, i.e. connecting one modem back-to-back with another. The external clock signal is then used to ensure a unique clock rate over the complete transmission path. The required circuits include a basic handshake on pins 4 and 5. This is used to allow half-duplex operation, though it is also one of many mechanisms used for flow control on full duplex circuits.

Additional features include auto-answering (pins 20 and 22), standby lower speed operation (pin 23), quality of detection of signals (pins 8 and 21) and a complete secondary channel just for control in a very few half duplex situations (pins 12, 13, 14, 16 and 19).

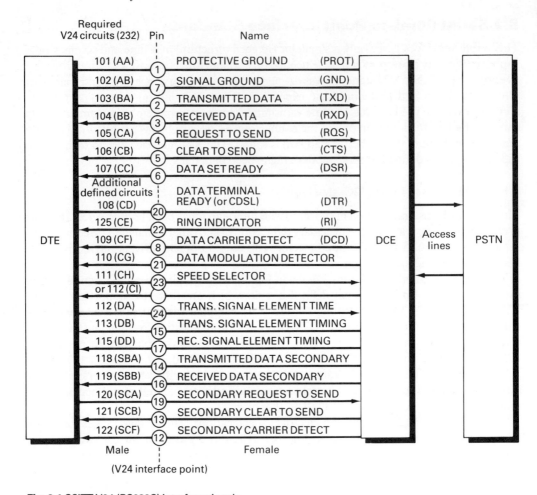

Fig. 8.1 CCITT V24 (RS232C) interface signals

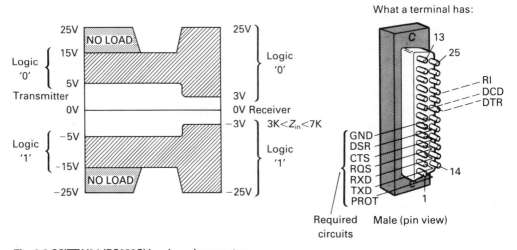

Fig. 8.2 CCITT V24 (RS232C) levels and connector

When connecting data terminal equipment i.e. terminals, computers, printers etc. to data communication equipment i.e. modems, the chosen set of signals for the required services are connected pin-to-pin directly. However if the interface is used to connect a terminal to a computer we are connecting DTE to DTE and must have a dummy DCE. Each of the DTEs will have a male connector on its back (or cable end) so our dummy DCE must have a female connector at either end. It is called a 'null modem' and crosses over the signals so that one DTE's transmitted data becomes the other DTE's received data and vice versa. Other signals are connected similarly. There are problems due to the lack of symmetry. Various signals are defined to originate in the DCE and have no counterpart in the DTE signals. If we are to use this dummy DCE then it must somehow conjure up these signals. Simple logic signals such as Data Carrier Detect (pin 8) can be connected to a fixed true level or a signal which approximates to its function. It could be supplied from either end e.g. Data Terminal

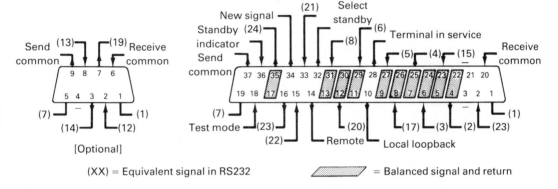

Fig. 8.3 RS449 circuit additions to RS232C

Ready (pin 20 also supplying Data Set Ready) looped back or cross-coupled. The difficulty comes with the clock signals discussed before. A box containing some logic and power supplies is needed to produce these signals so our dummy lead becomes a 'modemulator' to emulate two complete modems back to back. This is not a cheap or convenient solution just to connect two computers' synchronous ports together! The situation is made even worse by the manufacturers who insist on putting the wrong gender of connector on the back of their DTEs.

For the simple isochronous connection of terminals, computers, modems, printers etc. at speeds up to 10k bits per second this is a standard which is with us and it works but we could do so much better.

An attempt has been made to improve on the EIA RS232C standard. It is the EIA RS449 standard. To achieve the same ends (i.e. serial data transfer) for which it has been demonstrated 25 pins are more than is needed, the new standard uses a 37-pin connector. As if that were not enough an extra optional 9-pin connector is also defined. Figure 8.3 shows the additional circuit definitions over and above those which have direct counterparts in RS232C. Manufacturers do not seem to be falling over themselves in the rush to use the new standard. Perhaps they also regard it as the death throes of a dinosaur.

8.1.2 CCITT X21 standard

In the early 1970s the CCITT realized that their existing interfaces would prove inadequate for advanced wide area networks and proposed the X21 standard in 1972. It was finally agreed in 1976 and slightly amended in 1980. It was also a huge opportunity to replace the old V24 interface throughout the industry. Had the replacement been a success the new interface would now be on the back of every terminal and computer. It is not—even though X21 is simpler than V24 as services are provided at a logical rather than a physical level.

The X21 standard is for synchronous, full duplex, serial connection between data terminal and data communication equipment (DTE-DCE). A similar connection to V24 then fits onto a 15-pin 'D' connector and is shown in Fig. 8.4. The pin definitions and mechanical specifications are from ISO 4903. It is designed to be driven differentially with balanced drivers. This is to the X27 (EIA RS423) electrical specification. It was also arranged so that a lower speed mode (up to 9.6k baud was recommended) could use single-ended drivers with the same receivers (X26, RS422). The balanced mechanism uses two wires per signal with the pin allocation shown. There are six logical signals. Transmitted and Received data are the necessary two. As the interface is designed to be driven synchronously one would expect parallel clock signals for these two.

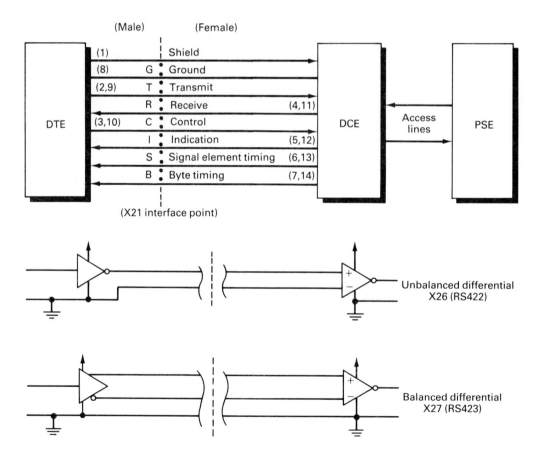

Fig. 8.4 X21 interface signals and drive arrangements

There is, however, only one data clock 'S', the Signal element timing. It is generated by the DCE and runs parallel with DTE Receive data. If the DTE wishes to transmit then it has to use the same clock. There is also a Byte time signal to indicate the last bit of a byte and thus define framing. It is not mandatory for the common carriers to implement this from their DCEs and so any data must always be prefixed by two 'SYN' characters to allow the devices to deduce the framing which will follow. All bytes are encoded for interface purposes from the alphabet number 5, which is the CCITT equivalent of ASCII. It is a seven-bit code to which an odd parity bit is added. The advantage of odd parity is that all zero and all one patterns are not legal.

Table 8.1 DTE and DCE signals for X21

DTE Signals 0, 1 logic levels, C=1=OFF, C=0=ON,<X>=ASCII characters

C	T	
1	0	Not ready uncontrolled
1	0101 (24 bits)	Not ready controlled (temporary)
1	1	Ready
0	1	Call Request or Accepted (in 500 ms)
0	<Syn><Syn><Address/Facility>	Selection Sequence
0	<Syn><Syn><Any data><Syn>	Data transfer
1	1	Terminate data (ready)
1	0	Clear Request or Confirmation (in 100 ms)

DCE Signals 0, 1 logic levels, I=1=OFF, I=0=ON, <X>=ASCII characters

I	R	
0	1	Not ready uncontrolled
1	1	Ready to enter operational phases
1	<Syn><Syn><+><+><+><+><+>	Accept selection (=dial tone)
1	1	Connection in progress
1	<Syn><Syn><XY>	Call progress signals (X96, 2 digits)
0	1	Ready for data
0	<Syn><Syn><Any data><Syn>	Terminate data
1	0	Clear Indication or Confirmation (2 s)

All control signals are arranged by using Indicate or Control. They are logic signals, rather than signals changing at data frequency, and are used to modify the meaning of the respective data lines. A simplified list of combinations of these and their meanings is shown in Table 8.1. Obviously sequences and timing constraints for permissible changes are included in the standard. The signal element timing is defined with less than 1% error and where two signals are changed, though the order does not matter, they must change within seven bit times. If the optional byte time signal is included this is reduced to a single bit time. Any logic level must persist for 24 bit times or until the other side responds to it.

During call establishment signals are returned to indicate the progress of the connection request. These range from the satisfactory response **through**, to one of busy, not obtainable, access barred, number changed, incompatible class of service, or not connected due to network congestion. Following a satisfactory connection the DCE indicates it is ready for data within two seconds of the last call progress signal. Test loopback may be set up remotely, automatically by the network and testing has priority over call establishment. A further problem can arise when call request is in 'collision' with an incoming call. The DCE detects this and abandons the incoming call so that it can proceed with the outgoing call. Many other facilities are provided for and examples of these are re-dial, line identification for both calling and called lines and the transfer of charging information.

The X21 standard as described seems to be quite elegant. It is clear and unambiguous and so we may ask why it has not rapidly taken over from V24. As it is intended to give interfaces of 48k baud or more over the wide area network and could run much faster locally, the question is even more pointed. I think the answer is that some of the fundamental rules for an interface have been broken: it is not symmetric and it allows optional choices as to how it is used.

The byte time signal, which is intended to give quicker synchronization, is quite pointless as even when it is included the cautious programmer will have to prefix his data with two SYN characters in case the 'B' line is not used at some point. The signal timing originates only from the DCE. The reason for this is presumably a hangover from the use of phase-shift keying to transmit higher-speed data over the analog telephone network. As the data was recovered differentially by comparing the current phase with that a symbol time ago it was necessary for the DCE to operate synchronously and to know what one symbol time was. No such restrictions apply to a digital network and so the DCE should take data at a rate determined by a DTE clock with no problems. If we wished to use the X21 interface to link two DTEs, for example personal computers, the lack of a DTE clock is our downfall. We cannot even make up a 'null modem' lead as is done for V24. If the 'B' signal had been omitted and a DTE 'S' had been defined then X21 would be the dominant serial interface now and for the future. As it is we are left still looking for something to replace all the 25 pin connectors on the backs of terminals.

8.2 Parallel Point-to-Point Standards

There are no widely used international standards for parallel point-to-point connection. One would have thought that a simple printer interface would have gained this standing, but only the 'Centronics' printer interface is very widely used. In many systems a single bussed interface is simpler and cheaper than a multiplicity of point-to-point connections. The 'Centronics' interface uses a 36 pin connector to carry eight data bits in parallel with strobe and status signals. It is a large and inconvenient connector, so much so that IBM have mapped the signals onto the 25-pin 'D'-type connector to give us another source of confusion in its use.

An attempt at a standard was made in 1969 with the ratification of the BS4421 interface. This is a unidirectional, parallel, point-to-point standard. It has eight data lines in parallel and an additional parity bit may be generated by hardware. There are two pairs of handshake signals. The first is a two-wire asynchronous transfer control mechanism as described in Section 7.2. The second pair indicate the operability of the two ends, source and acceptor. Also provided are an error signal for the receiver to indicate incorrect parity back to the source, and a source terminate signal. Once the 'operable' signals are both raised data is transferred by the asynchronous control mechanism at a rate determined by the cable delays and the receiver. The standard is simple and neat but failed to gain popularity for three main reasons. Firstly, two different connector types were permitted. Secondly the voltage levels chosen were ±12V like V24 and this required a complexity of drivers which 0V/5V signalling would have saved. Finally it was simply not taken up by any major manufacturer and so remains another standard definition gathering dust on the bookshelf.

8.3 Standard Bussed Interfaces

There has been a number of successful bus interface standards probably because of their general purpose nature. As has been discussed before they split into two types: peripheral buses and backplane buses. The main difficulty of standards designers (committees) has been to know how far to go. Define too little and the standard fails because devices may be incompatible, but define too much and the standard fails because it is too restrictive. We will see examples of both of these in agreed international standards, but we will start with one where a balance appears to have been achieved.

8.3.1 IEEE 488 (the GPIB)

The Hewlett Packard Corporation first published their approach to the design of an instrumentation interface in 1972 and subsequently the full specification of the interface bus. HP was renowned for an extensive range of measuring equipment and needed a way of linking multiple devices under computer control to give extremely flexible systems. The HPIB resulted and was standardized by adoption as the 'Standard Digital Interface for Programmable Instrumentation' IEEE 488 (1975). It was also accepted as ANSI MC1-1975 and is sometimes called the general-purpose interface bus (GPIB) presumably by manufacturers not wishing to use the name of another company. The standard was revised in 1978 to incorporate some minor cosmetic changes.

The bus may be up to twenty metres long and can incorporate fifteen devices. It operates fully asynchronously and may reach a data rate of 1 M bytes per second. There are three classes of device specified: **listeners**, **talkers** and **controllers** which are shown connected to the bus in Fig. 8.5. There are eight bidirectional data lines, three data transfer control lines and five general management lines, with six 'fast' signals having separate twisted pair return wires. Adding an overall screen (shield) and a common signal return line this gives a 24-pin connector with the layout shown. The problem of having the wrong 'sex' of connector on a lead is solved simply, at the same time easing the connection of a daisy chain cable layout. Standard leads are made up with both male and female on each end in a single moulded unit.

The voltage range for drivers and receivers is shown in Fig. 8.6. It fits with the ordinary 5V supply but needs more current drive than is usually available from a micro or interface chip. Additional drivers are normally used and are often found combined in two ICs. As commonly used all the lines are connected to open collector outputs with parallel termination and pull-up resistor to a 5V supply. A negative logic convention is used throughout the bus so that signals are true at zero volts. These factors meant that transitions from true to false were slower than those from false to true due to the passive pull-up and the active pull-down arrangement. The later use of tristate drivers enables maximum speed to be reached more easily.

It should be noted that return wires for the 'fast' signals do not include the data lines. The data transfer control signals are the really fast signals as they implement the three-wire, asynchronous, multipoint data transfer regime described in Section 7.2. This is shown with the GPIB nomenclature in Fig. 8.7. Firstly the currently assigned listening devices all assert that they are ready to take data. The last one to become ready actually lets the pull-up resistor pull up the signal to approach five volts. So 'Not Ready For Data' is set false. The currently assigned talker (there will, or rather should, be only one) then puts data on the eight data lines and indicates its validity by pulling 'DAta Valid' true, close to zero volts. The listeners then release their 'Not Ready For Data' outputs to true and when they have read the data, set 'Not Data ACcepted' false. These two events can happen at the same

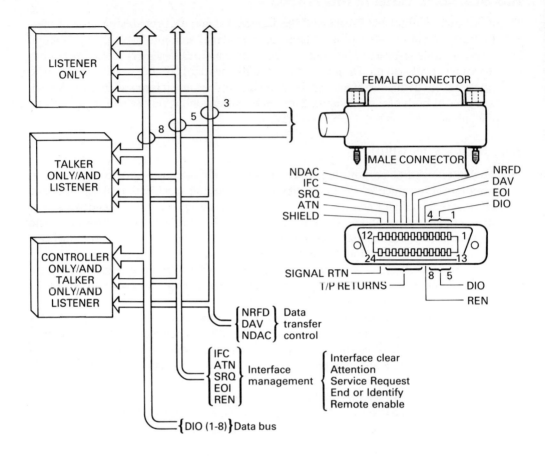

Fig. 8.5 IEEE 488 interface—layout and signals

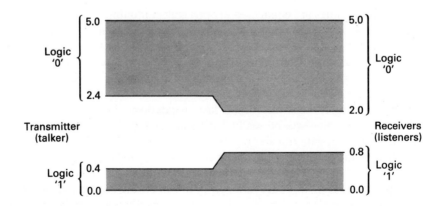

Fig. 8.6 IEEE 488 interface—logic levels

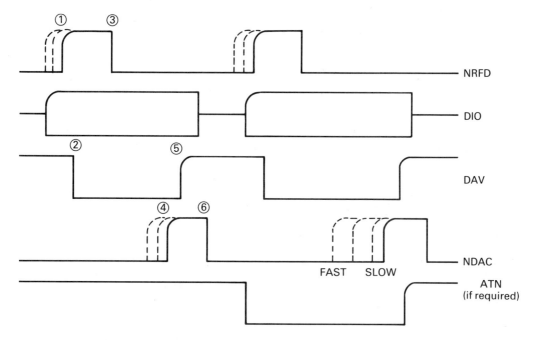

NRFD

DIO

DAV

NDAC

FAST   SLOW

ATN
(if required)

Fig. 8.7 IEEE 488 interface—bus handshake control

time. Again the last one to release the line lets it be pulled up to tell the talker that the data has been taken by all the devices. The talker then releases data valid to let it be pulled up (false) and removes the data from the data lines. The listeners then set their not data accepted back true prior to again asserting that they are ready for the next transfer. Devices carry out a number of checks for the possible error conditions in the transfer process.

So the basic data transfer mechanism is simple and automatically adjusts to the data rate of the slowest device, thus giving a reliable system. It is at the next higher level where the power of the GPIB becomes really apparent. The 'ATN' line is used by the current controller device to inform devices on the bus that the transfer is of control information which may affect them. If ATN is false then the transfer is of data only. The main functions of a controller are to send interface messages for establishing addressing and polling configurations, to receive and pass on control, to respond to service requests (SRQ) and arrange serial or parallel polling and initialize the interface (IFC). In addition it may clear devices, lock out or enable their local controls and configure them to operate locally or under remote control (REN), and trigger a group of devices to give synchronized, co-operative action.

When the controller sets ATN to be true the current talker, if any, relinquishes its control of DAV even if only temporarily. The controller can then carry out any of its command functions, the primary ones being shown in Fig. 8.8. There are a group of **universal** commands which are acted on by all devices on the bus and a group of **addressed** commands. Perhaps we should look at how addressing works on the bus. Any device able to be a listener has a listen address. Similarly for the talkers. To remove all listeners from the bus the **UNListen** command is sent. To configure a group of listeners we simply issue a series of appropriate listen commands from the controller. To remove any existing talker from the bus the **UNTalk** command is sent. To configure the chosen talker we send its talk address

ATN = TRUE	7	6	5	4	3	2	1	0	Data DIO
BUS COMMAND*	X	0	0	C	C	C	C	C	Codes below
LISTEN ADDRESS	X	0	1	L	L	L	L	L	
UN-LISTEN	X	0	1	1	1	1	1	1	Remove all listeners
TALK ADDRESS	X	1	0	T	T	T	T	T	
UN-TALK	X	1	0	1	1	1	1	1	Remove current talker
SECONDARY ADDRESS/ COMMAND	X	1	1	S	S	S	S	S	See text

* CODE	UNIVERSAL COMMAND		* CODE	ADDRESSED COMMAND
11	LLO LOCAL LOCKOUT		01	GTL GO TO LOCAL
14	DCL DEVICE CLEAR		04	SDC SELECTIVE DEVICE CLEAR
15	PPE PARALLEL POLL UNCONFIGURE		05	PPC PARALLEL POLL CONFIGURE
18	SPE SERIAL POLL ENABLE		08	GET GROUP EXECUTE TRIGGER
19	SPD SERIAL POLL DISABLE		09	TCT TAKE CONTROL

Fig. 8.8 IEEE 488 command mode code assignment

command while ATN is true. This also has the effect of removing any other talker from this role as there may only be one active talker at any time. The device's own addresses are set in switches or loaded into single chip versions of the interface by the micro to which they are attached. Secondary addresses are used to select sub-units within a device.

The first universal command allows the front panel controls of all connected devices to be disabled or locked out. The device clear command returns all devices to their predetermined clear state. Parallel poll unconfigure clears all devices with parallel polling response capability to the not responding condition. This, with the two serial poll commands, is discussed with the service request mechanism further on.

Addressed commands are only acted on by devices currently configured as listeners. 'GoTo Local' returns the device(s) to local control and permits operation of any front panel controls. The selective device clear returns only the current group of listeners to their predetermined clear state. There is an explicit mechanism for transferring control around among controlling devices. This has been discussed in Chapter 7 but is represented here by the 'Take Control' command. Obviously only one controller device should be chosen as a listener prior to the issue of this command. Also, one controller device must be specified as the system controller and it has control of the Interface Clear (IFC) and Remote Enable (REN) signal lines. These permit an orderly start-up of the bus and recovery in case of serious failure. Finally comes the most important command for combining measurement devices and one which betrays the origins of the IEEE 488 bus. On an ordinary bus if a number of devices are to be started up then sequential commands are issued. This results in their starting at slightly different times. Using the Group Execute Trigger a sequential set of setup commands can then be activated simultaneously. An analogy is a parade ground where the Sergeant Major prepares the troop with the Atten. . . and group-executes with the . .Shun!

The secondary group of addresses/commands allows an extended range of talker or listener addresses within a device. They are also used to configure the parallel poll as below.

So we can set up devices to receive or transmit and then let them carry out a block of transfers. The End-Or-Identify signal (with the ATN signal false) is used to show the end of such a block. We can control and clear devices and move control around. It remains to describe the interrupt mechanism which is needed in any multipoint interface.

If a device needs service then it sets the Service Request (SRQ) true. This allows a device to gather or disperse data over a period of time and **request** controller action only when it is ready for a transfer. When this is detected by the current controller it has to determine which device (or devices) are requesting. There are two mechanisms: one fast but restricted, the other slower but universal. The active controller can poll all the devices in series to ascertain which device(s) request service.

To carry out this serial poll the controller unlistens all devices and then issues the Serial Poll Enable (SPE) command. It then addresses the first device as a talker, the previous actions all being carried out with ATN true. When ATN goes false the talker sends a one-byte status word. Bit 7 is set true to indicate if it is requesting service and the other bits are available for device-dependent status bits.

The controller then addresses the next device (ATN true and a talker address) and receives its status (ATN false, device sends status). When the poll sequence is completed to the controller's satisfaction it terminates the process by issuing the Serial Poll Disable (SPD) command. It then issues another talker address or issues UNTalk to remove the last polled talker from the bus. This complete process is obviously slow and so a much faster method is included but it can only operate for eight devices. As there are eight data lines we could allocate one per device so that a single command will retrieve a single status bit from each device. This would locate the requesting devices and getting any more detailed status would only need a single serial poll.

To set up a parallel poll mechanism the controller issues a parallel Poll Unconfigure (PPU) to remove any previous configuration and then addresses the first device as a listener. It then issues the Parallel Poll Configure (PPC) command and follows this with a secondary command. There are five 'S' bits in this as shown in Fig. 8.8. The top 'S' bit (bit 5) is a '1' to disable parallel poll on this device (PPD). If the top bit is a zero then the next bit indicates the sense of the response (0 or 1) and the least significant three bits indicate on which of the eight data lines this device will respond. The setup for this device is finished by the UNListen to remove it from listening to the poll configuration. Once all devices to be parallel polled have been configured then the controller can call for this at any time.

The controller simply sets both ATN and EOI true to force all parallel poll devices to respond with their 'request' bit on the appropriate data line. This mechanism is then very fast, taking only a single cycle with no handshake needed. It is simple but has the limitation of eight data lines so only eight devices can use the parallel poll for any given period.

The IEEE 488 is representative of very well-designed interfaces. It is complete but very flexible in use. There is one curious anomaly caused by the multitude of standards-making bodies. When the International Electrotechnical Commission agreed their IEC 625-1 standard it was identical to the GPIB in almost every respect. For some reason they chose to use the ubiquitous 25-pin 'D' connector but with male and female pairs at each end of the leads. The extra pin is an earth. The Remote Enable (REN) line is moved to between DI04 and EOI. All the others move up one pin so the shield goes on pin 13. There are then eight grounds (pins 18–25) and each is twisted with the signal on the pin opposite, i.e. every signal except the eight data lines is a twisted pair.

8.3.2 IEEE 583 (CAMAC)

The Computer Automated Measurement and Control standard (CAMAC) was first proposed in 1961 and its ideas were mainly established by 1964 in Harwell (UK). In 1967 responsibility for development of the standard was accepted by the ESONE committee of European Scientific Laboratories. It was revised and accepted by the US Nuclear Instrument Module (NIM) committee in 1970 and subsequently was adopted as IEEE 583. It was designed for use with nuclear physics instrumentation and hierarchical computer architectures and consequently is not so well suited to process control. CAMAC was the first successful attempt at a complete system interface design including all aspects of electrical, functional and mechanical characteristics. The main specification starts with a physical 'crate' which is a twenty-five-slot card rack with a 'Mother' card interconnecting them, with a power supply mounted in it. The crate must include a crate controller card and an interface card to connect to a computer. If these functions are combined, possibly even with the computer itself, the two slots at the right-hand end are still needed to get access to all the lines. Because a complete crate with controller and full power supply are essential even for a single card, CAMAC systems are expensive. But because the specification is so complete one can build up a complex interface by buying and plugging in.

The power supply must supply ±6V and ±24V. In addition provision must be made to include all the optional supplies for which lines also run to each card slot in the crate. These supplies are ±12V, +200V DC and 117V AC! The fourteen power lines and the 72 signal lines are run along the back of the crate as a rather complex bus structure called the **Dataway.** There are twenty-four **read** data lines and twenty-four **write** data lines. The use of unidirectional lines in the CAMAC fashion means that modules cannot talk to each other, or the computer, without using the controller as an intermediary transfer buffer. Addressing is rather complex. Modules do not contain an address but are addressed by their slot position. Thus they cannot be moved without corresponding changes to the software. There can be up to eight crates in a system connected in ways which will be discussed further on. There are 23 slots left for interface modules in each crate as has been shown. Each module may have up to 16 sub-addresses (4 lines). Thus a complete address issued by the computer is:

$$<Crate + module\ Number + Subaddress>,$$

with a Function to be carried out (i.e. CNAF). Within a crate the controller decodes the five-bit number to an individual select line which runs to each slot separately. The functions are carried on five lines and are summarized in Table 8.2. There are also three separate control lines used for functional purposes.

Initialize (Z) sets all registers and control functions of the modules in a crate to a defined initial state. This is normally done on powerup or if an error condition is detected. The clear (C) signal resets registers and bistables of modules during normal running. The inhibit (I) disables features for the duration of the signal.

Table 8.2 CAMAC functions

000 (Read)	100 (Write)
00 Read Register	00 Overwrite Register
01 Read Register (Group 2)	01 Write Register (Group 2)
10 Read and Clear Reg	10 Selective Write Register
11 Read Complement of Reg	11 Selective Write (Grp 2)
010 (Clear)	110 (Misc)
00 Test Look At Me	00 Disable
01 Clear Register	01 Increment Register(s)
10 Clear Look At Me	10 Enable
11 Clear Register (Group 2)	11 Test Status

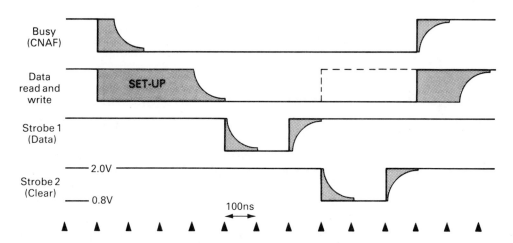

Fig. 8.9 CAMAC dataway timing

Timing and data transfer on the CAMAC dataway is most interesting as it is a fully synchronous system and any function takes one microsecond. Figure 8.9 shows the timing for set-up and the subsequent two strobes. The first strobe does read/write transfers whilst the second is normally used for clearing registers so that a dual function need take only one cycle. For a system defined so long ago CAMAC is surprisingly fast. However, it cannot cope with other than the fixed timing. Once a command is received by a module and the module commences to act upon it, the module asserts the command accepted (X) line. This allows the system to determine if a module exists in a given place and can perform a given command. Provision has been made for block transfers to run continuously at high speed by setting up a command and repeating it using the Q-response line to indicate the end of the block.

The final interfacing signal is 'look at me' (LAM) which is an interrupt line running separately from each slot back to the controller. There it is held along with the others in a 24-bit register and with a parallel mask register. The use of masking and AM grading allows a flexible priority interrupt to be arranged. Issuing N(30), A(0), F(0) lets the computer read the interrupt (LAM) status.

Multiple crates are connected together in two ways. These are called the CAMAC Serial

Highway and the CAMAC Branch Highway. The serial highway consists of two twisted pairs which carry all the necessary signals in a serialized form. This is cheap for long distances but slows the overall operation markedly. For more local connection the branch highway is a bundle of 66 twisted pairs providing an effective extension to the common dataway signals with the added crate number.

CAMAC is a rigorously specified interface, in fact, more of a 'whole system concept' than just an interface. It includes formatting and software conventions to attempt to ensure that such a complex arrangement is simple in use. Unfortunately it is far too narrow a specification to be used as a universal interface. It is a hierarchical system normally only for a single computer. It is possible to interface two or three computers to a master crate but it is difficult to organize the control. CAMAC is thus at odds with modern distributed approaches and equal status interfacing. The starting cost of CAMAC is very high by comparison with IEEE 488 for example. There is, however, a very wide range of modules available ranging from D-As and A-Ds to scalers, counter/timers to peripheral controllers, yet it is a complex system needing quite a lot of software support and has a steep learning curve. Despite this it has a momentum all of its own in experimental physics establishments, but is not used for modern control or logging interfaces.

8.4 Backplane Bus Standards

Inside a boxed computer there is a need to connect various printed circuit boards to make up the machine and to allow optional additions. Such interconnection is usually handled by a parallel run (or backplane bus) of tracks between the sockets on a 'mother-board'.

There are three generic problems with most backplane buses. Firstly, there are far too many international and 'industry' standards in this area. Secondly, their designers have often failed to understand the fundamental purpose and functions of a backplane bus. These two have led to the standardization of elementary mistakes. Finally, most backplane buses have been designed with very short time horizons. They have not allowed for even modest expansions in address range, data size, or bus control etc. They have been designed, in many cases, with a specific processor or a specific operating paradigm in mind, even when such restriction was quite unnecessary. This, then, is why there are so many standards in this area. Each manufacturer has tried to impose his own ideas and so has embodied them in a range of products. Such an approach is restrictive for the user community.

The most successful standards have come when a well-designed bus has been placed in the public domain, free from patent, copyright and trademark limitations.

No purpose-designed generic or automatically sized bus has been standardized yet, so buses are normally considered to be sized by their **data** width. It happens that such a categorization also correlates well with the address range, power and performance of systems built on such buses.

Historically, the first standard eight-bit backplane bus for microprocessor use was the **S100** bus. It was designed by the Altair Corporation to support the Intel 8080 processor, and was released in their MITS small computers in 1975. It was latched onto by many other manufacturers and became, after some serious contortions, the IEEE 696 standard in 1982. By that time its fundamental weaknesses had rendered it obsolete for any new designs. But, as it is recognized as the first of the microcomputer buses, it does warrant some further study. A comparison of the signal and power supply arrangements of the S100, STD, STE, and 32-bit VME busses is shown in Figs 8.10 to 8.12. When first used, S100 had uni-directional data lines: eight for input data and eight for data output. It had sixteen address

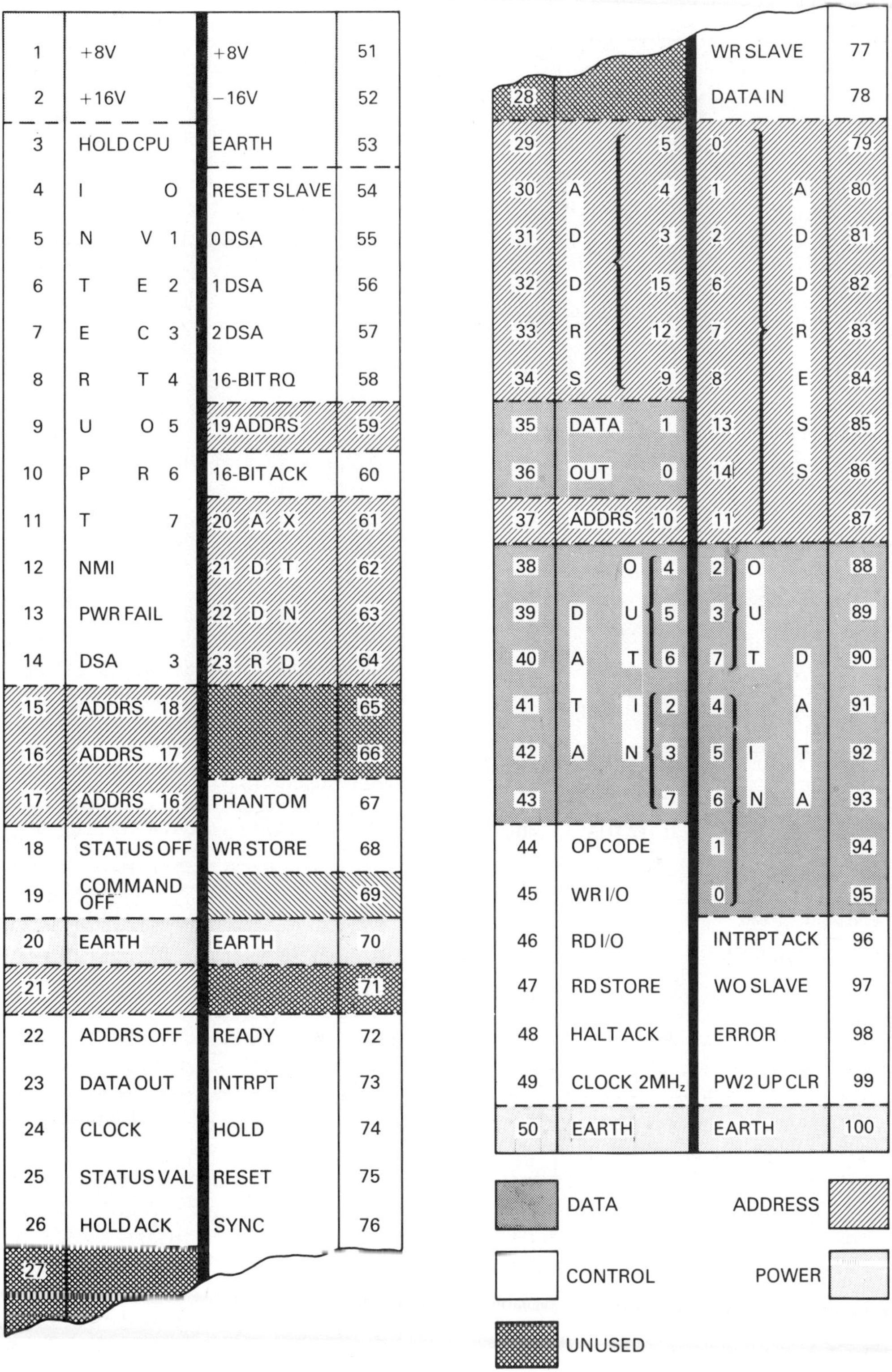

Pin	Signal	Signal	Pin
1	+8V	+8V	51
2	+16V	−16V	52
3	HOLD CPU	EARTH	53
4	I O	RESET SLAVE	54
5	N V 1	0 DSA	55
6	T E 2	1 DSA	56
7	E C 3	2 DSA	57
8	R T 4	16-BIT RQ	58
9	U O 5	19 ADDRS	59
10	P R 6	16-BIT ACK	60
11	T 7	20 A X	61
12	NMI	21 D T	62
13	PWR FAIL	22 D N	63
14	DSA 3	23 R D	64
15	ADDRS 18		65
16	ADDRS 17		66
17	ADDRS 16	PHANTOM	67
18	STATUS OFF	WR STORE	68
19	COMMAND OFF		69
20	EARTH	EARTH	70
21			71
22	ADDRS OFF	READY	72
23	DATA OUT	INTRPT	73
24	CLOCK	HOLD	74
25	STATUS VAL	RESET	75
26	HOLD ACK	SYNC	76
27			
28		WR SLAVE	77
		DATA IN	78
29	A D D R S 5 0 4 3 15 12 9	D A T A 0 1 2 6 7 8 13 14 11	79
30	A	A	80
31	D	D	81
32	D	D	82
33	R	R	83
34	S	E	84
35	DATA 1	S	85
36	OUT 0	S	86
37	ADDRS 10		87
38	O 4	2 O	88
39	D U 5	3 U	89
40	A T 6	7 T D	90
41	T I 2	4 A	91
42	A N 3	5 I T	92
43	7	6 N A	93
44	OP CODE	1	94
45	WR I/O	0	95
46	RD I/O	INTRPT ACK	96
47	RD STORE	WO SLAVE	97
48	HALT ACK	ERROR	98
49	CLOCK 2MHz	PW2 UP CLR	99
50	EARTH	EARTH	100

Legend: DATA, ADDRESS, CONTROL, POWER, UNUSED

Fig. 8.10 S100 bus IEEE 696

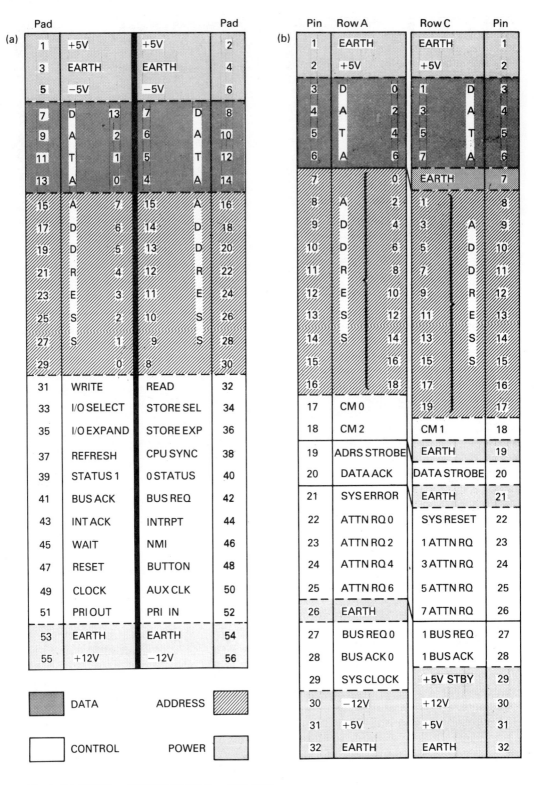

Fig. 8.11 (a) STD bus IEEE 961 (b) STE bus IEEE 1000

lines and with eight interrupt vector lines, an amazing thirty-nine other control signals and six power lines left fifteen undefined and unused connectors. The large number of control signals was caused by the peculiar requirements of the very early Intel microprocessors. Following standardization, extra control signals were included to allow data to be transferred as either eight-bit unidirectional or sixteen-bit bidirectional. An extra eight address lines were added and the direct store access arrangement was enlarged. The power supply remained the same, though, and only unregulated DC voltages are supplied. Raw plus eight volts and plus and minus sixteen volts are connected onto adjacent pads. This rather spoils the claimed advantages of individual regulators on each board spreading the heat dissipation and limiting failures to a single card. The experiment of distributing unregulated DC has not been a success. Central, switching-mode regulators are more efficient, give better regulation and allow simple power-fail detection circuits to give a full auto-restart capability. The use of printed circuit 'finger' contacts, even when gold plated, is not fully reliable either. Full pin and socket **indirect** connectors are a much better solution and almost all of the advanced buses use them. The S100 card-edge connector has a further problem. It can be plugged in the wrong way round! The disaster this causes can be seen from the pad allocation shown in Fig. 8.10. Obviously this problem should be avoided by using asymmetric connectors. If a card-edge connector is inserted or removed with power supply on then adjacent pads can short together, again causing damage.

Not long after S100 was released another backplane appeared, to remove some of the problems. It too evolved within a manufacturer and was standardized rather too late. The **Multibus** was originally used by the Intel Corporation for their SBC 80/10 single-board computer and its derivatives. It was finally standardized as IEEE 796 just as the uprated, but incompatible, Multibus II came into use. The 796 bus is defined as a double-sided, gold plated, printed circuit card-edge connector like S100, but only needs forty-three pads on each side. There is an additional 60-pin auxiliary connector designated completely for user signals, and this gives asymmetry for insertion.

The process of standardization increased the Multibus I address bus from sixteen to twenty lines, added an interrupt acknowledge signal, but lost the -5V DC and 10V analog reference power lines. So the IEEE 796 standard is for sixteen-bit, bidirectional data and a one-megabyte addressing range. Both 696 and 796 can provide for sixteen-bit data paths and twenty bits of address. Both are somewhat more complex than is needed and are often used to support eight-bit microprocessors. The next two standards, shown in Fig. 8.11, were designed specifically for a byte width data path. Instead of a hundred or even eighty-six pads, only fifty-six pads for STD (standard) and sixty-four pins for STE (standard euro) are needed. Both represent simple, well-designed backplanes which support all functions without unnecessary frills.

The STD bus, designed by the Prolog corporation, was released into the public domain in 1978, and continues as IEEE 961. It uses a fifty-six way double-sided, gold-plated, card-edge connector with the pads on a one eighth inch pitch. The card is a little larger than the single Eurocard standard, being about 114 mm by 165 mm. A small system including processor, store and I/O can be fitted on a single card. Expansions usually also fit on a single card as they are made from eight-bit data path chips. The small card size gives resistance to vibration and shock in harsh environments. It is also convenient for portable equipment. If the sixteen-bit addressing range proves to be inadequate there is an extension signal to allow bank switching to give multiple address spaces. Each new bank gives up to 48 kilobytes extra, with a 16 kilobyte common store area at the top of store for interprocess communication. The interface control mechanisms are simple. A bus request/acknowledge system handles mastership of the bus, and a daisy chain gives interrupt priorities.

Having accepted the desirability of a small-card-based, eight-bit data bus system, there still remain restrictions inherent in the STD bus. Bank switching may not be good enough as a way of increasing the address range. Though it works well enough for traditional styles of program access which have known and fairly regular transitions from bank to bank, data accesses are not well catered for. If the effective address of a data item is computed at run-time then a bank switch may well be required during execution of an instruction. More address bits are the only solution. For the STE bus a single Eurocard format (100 mm × 160 mm) was chosen with indirect IEC-603-2 (DIN 41612) standard connectors. These have two rows of 32 pins each. Four of the extra pins are used to extend the address range to one megabyte (20 bits). Asynchronous data transfer control combined with bus-mastership transfer by dual bus request and bus grant lines allow for multi-processor configurations and system expansion. This is also catered for by permitting an extension to the basic board's form factor to double Eurocard size. This is to accommodate any high voltage or high current I/O connections at the bottom of a rack, the bus connector always residing at the top. Physically, the backplane design and pin layout are intended to minimize cross-talk by the judicious use of earth tracks. The pin layout is shown in Fig. 8.11b, and general signal layout is discussed in detail in Chapter 10. Logically, the asynchronous data transfer control is extended to give both **read-modify-write** and **burst** data blocks. These are essential for semaphore handling in multi-processor systems and to achieve high data rates when needed. Again, both bus release when transfers are complete, and bus release on request are supported to match specific device requirements. The STE bus (IEEE 1000) is a well-designed solution to the backplane bus problem for any eight-bit organized processors. As most manufacturers offer eight-bit data bus versions of their microprocessors, STE has a wide range of application. It can even be used as an I/O bus within a larger, more powerful system.

As the power of processors has increased so the demands placed on the backplane bus communication have increased. This will eventually reach the point where backplanes will no longer be able to cope and intercommunication will have to be provided by networked point-to-point paths. Though there are intermediate numbers the next obvious 'size' can cope with 32-bit data path, 32-bit address range systems. There is a number of very basic choices for 'high performance' buses, but after they have been taken there are some remarkably similar specifications for standards. This is a shame as minor differences are really no reason for a multiplicity of standards. Examples of buses in the high-performance category are the VME bus (IEEE 1014, IEC 821), Multibus II sponsored by the Intel corporation, Fastbus (IEEE/ANSI 960) and the Futurebus (IEEE 896). The first of these to be released was the VME bus by the Motorola corporation, and it represents a staged approach.

The primary connection to the bus, shown in Fig. 8.12, only contains 24 address lines and 16 data lines. This means that smaller systems may be constructed using only single Eurocard sized boards, like STE but having 16-bit wide data transfer paths. An additional connector, making the board double Euro sized, contains the extra eight address lines and sixteen data lines for full-sized systems. This additional connector also contains 64 uncommitted I/O lines for users to allocate as they wish. This approach always **reduces** flexibility as it is unlikely that all boards will use them for the same purposes. In 1984, the Intel corporation released their Multibus II specification to get over the shortcomings of Multibus I. This specification actually defines no less than five separate communication buses and attempts an overall 'system' definition. The main bus is a parallel system bus. There is an additional serial bus, similar in role to the VME serial lines. A local bus extension provides

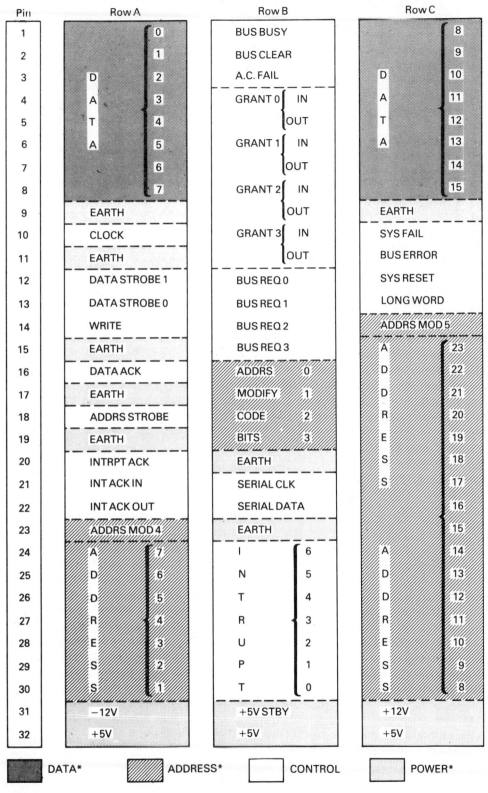

Pin	Row A		Row B		Row C	
1		0	BUS BUSY			8
2		1	BUS CLEAR			9
3	D	2	A.C. FAIL		D	10
4	A	3	GRANT 0	IN	A	11
5	T	4		OUT	T	12
6	A	5	GRANT 1	IN	A	13
7		6		OUT		14
8		7	GRANT 2	IN		15
9	EARTH			OUT	EARTH	
10	CLOCK		GRANT 3	IN	SYS FAIL	
11	EARTH			OUT	BUS ERROR	
12	DATA STROBE 1		BUS REQ 0		SYS RESET	
13	DATA STROBE 0		BUS REQ 1		LONG WORD	
14	WRITE		BUS REQ 2		ADDRS MOD 5	
15	EARTH		BUS REQ 3		A	23
16	DATA ACK		ADDRS	0	D	22
17	EARTH		MODIFY	1	D	21
18	ADDRS STROBE		CODE	2	R	20
19	EARTH		BITS	3	E	19
20	INTRPT ACK		EARTH		S	18
21	INT ACK IN		SERIAL CLK		S	17
22	INT ACK OUT		SERIAL DATA			16
23	ADDRS MOD 4		EARTH			15
24	A	7	I	6	A	14
25	D	6	N	5	D	13
26	D	5	T	4	D	12
27	R	4	R	3	R	11
28	E	3	U	2	E	10
29	S	2	P	1	S	9
30	S	1	T	0	S	8
31	−12V		+5V STBY		+12V	
32	+5V		+5V		+5V	

DATA*   ADDRESS*   CONTROL   POWER*

*See text

Fig. 8.12 VME bus, IEEE 1014, IEC 821

for increased bandwidth between two adjacent boards. A connector for piggy-back extensions on a main card, and the direct store access multichannel I/O bus are kept on from the original Multibus I specification. The main Multibus II is a synchronous transfer system with multiplexed address and data lines. It can perform single transfers at 20 megabytes per second and reach burst rates of 40 Mby/s, giving almost identical performance to the non-multiplexed VME bus.

The main drawbacks of Multibus II are caused by its synchronous physical transfer mechanism. Though its transfers are logically asynchronous the performance of the bus is forever limited to sub-multiples of the clock frequency, peaking at 40 Mby/s. Also the bus is limited by this to 16.8 inches in length and so passive card extenders may not be used to ease testing or fault finding. Fully populated, maximum-length systems are unlikely to achieve the maximum rates and would have to operate at sub-multiples of the clock rate. This is because the drivers only provide 64 mA maximum drive current and so the voltage swings are not adequate at full load. The VME bus has its own, smaller, set of problems, but both were designed and released and only standardized later.

Both Fastbus and Futurebus were not manufacturer's own interconnection buses which were then released, but were designed by user organizations. Fastbus came from the US Nuclear Instrument Module Committee (NIM) while Futurebus is the work of the IEEE 896 working group themselves, and is the most adventurous approach. Both use different driving techniques to get over the loading problem of matching drivers to the characteristic impedances of the transmission paths. Fastbus uses emitter coupled logic (ECL 10K) which with its smaller voltage swings and constant output impedance is a good solution to the problem. Futurebus uses new 50 mA backplane transceiver logic, with narrow voltage swings, to give correct matching and a very good solution to the problem. Both are then designed to be asynchronous so that they can transfer data faster as systems get faster. They will cope with rates in excess of 100 megabytes per second. Even though in theory synchronous transfers should be faster, at very high transfer rates the delay along the bus is critical so that transfers between adjacent cards could run faster than those between the ends of the bus. So the asynchronous regime is a much better choice as it optimizes performance for **all** transfers. To cut down on the number of pins, and drivers and receivers of course, these buses multiplex the data and address signals onto the same pins. This allows Futurebus, for example, to use just one three-row (96-pin) IEC-603 connector for all its signals. This is the same connector on which VME fits its reduced set. There are obviously a lot more details of the high-performance buses which could be presented, but many are rather esoteric. However, Futurebus represents the current state-of-the-art in high-performance bus design.

The future for any high-performance bus is very dependent on the availability of silicon support chips for the drivers, receivers, arbitration and protocol logic. For any really detailed consideration of standard interfaces it is the **standard** which should be read, rather than a brief description or comparison as presented here. Though the standards necessarily make heavy reading, they are the only places which can be trusted for explicit, correct answers.

9

Output Transducers

If we want to have any effect on the world outside the computer we must have output transducers. They are the means by which we actually control anything. There are fewer output transducer types than input ones as we have a number of common mechanisms for adjusting a variety of variables. As with input transducers some categorization is helpful but, as we use common methods for many types of output, the split into different variables is less important. There is no equivalent to the passive/active split of input transducers as all output transducers require a power source whether it is electrical or fluid pressure. There is a fundamental split of the output techniques used though:

Output technique

(a) **Direct**	(a) **Digital**	(a) **Incremental**
(b) **Indirect**	(b) **Analog**	(b) **Absolute**

Direct-output driving achieves its aims by some form of **switching**. The output can be altered by a change of current (or voltage) flowing in a device. Heat and light intensity are obvious examples of this. Indirect outputs need some intermediate movement, normally called **actuation**, to effect the change. In this way we can generate sound, open or close valves, move or stir things. The variables we have to input, listed in Table 2.1, are conveniently split up into these two groups and those which may use either technique.

Direct	**Either**	**Indirect**
Electrical current	Composition	Acceleration, Mass
Frequency	Energy	Displacement, Pressure
Luminous Intensity	Power	Flow, Sound, Strain
Magnetic Flux	Temperature	Force, Velocity
	Time	Humidity, Vibration

The techniques may be further divided by the methods employed. For direct-output switching we have:

(a) **DC switching**	(a) **On/off switching ('bang/bang' control)**
(b) **AC switching**	(b) **Linear control**
	(c) **Phase control**
	(d) **Zero-voltage switching**

The load we are switching may use alternating or direct current. We may control it by just turning it on or off, or to some intermediate values. This may be done by controlling its amplitude or the time for which the current is applied to the load. Finally the time at which we switch may be important. When we look at indirect-output transducers we find that there are three possibilities for the prime motive power of the actuator. It should be noted that all must start off as a switched electric current even if this is only used as a signal to control the prime mover.

(a) **Electromagnetic**   (a) **Rotary movement**
(b) **Pneumatic**   (b) **Linear movement**
(c) **Hydraulic**

Electromagnetic drive may produce linear movement using solenoids or rotary movement by AC, DC or stepper motors. Fluid drives may produce linear movement using pistons or rotary movement using vanes mounted axially on a shaft. A combination of these can be arranged to produce an electro-hydraulic pulse motor.

9.1 Output Driving and Switching

The simplest output switch is a transistor. We already have them on the output of any integrated circuit (IC), probably in one of the arrangements shown in Fig. 7.11. The output current of standard integrated circuits is low to keep the required heat dissipation low. The voltage is limited either to the normal supply or, if an extra supply is used, to about 30 V. The IC transistors break down above this. If 5 V 1 mA switching will do the micro can drive it directly. For 1 mA to 50 m and up to 30 V an additional buffer chip can cope. An external transistor can be used if we need more. There are some small points about direct switching with transistors. Figure 9.1 shows an open collector transistor with various loads. The simplest is resistive with a linear voltage/current relationship. It is of use as a heater if resistive wire is used. The power available is low but there are some applications where such a small heater is of use. The second circuit has a light-emitting diode as the load. The voltage drop across a LED is about 2 V so we must include a current-limiting resistor to drop the rest of the voltage from the power supply. The resistor value is fixed to set the current by:

$$RI = V_{cc} - V_{led} - V_{ce}$$

where V_{cc} is the supply and V_{ce} is the collector–emitter voltage of the transistor when saturated, i.e. turned on! If we wished to drive an inductor or coil then it would appear that the circuit of Fig. 9.1c is all that is needed.

A coil when energized by a current passing through it generates a magnetic field. The right- hand rule (of thumb) shows how in Fig. 9.2. This magnetic field is a store of energy. This principle was used for magnetic-core storage which used to be the dominant real-store medium. When the current is turned off all the excess energy of the magnetic field has to go somewhere. The process reverses and the collapsing field generates a current in the coil in the **reverse direction**. The current causes a voltage often termed a back e.m.f. (electro-motive force) which would reverse bias the transistor and destroy it. To prevent this damage a diode is included as in Fig. 9.3a.

Unfortunately we have still not provided anywhere for the energy to go other than in heating the coil and diode. This makes turning the coil off quite slow. Including a resistor with the reverse (protection) diode in Fig. 9.3b solves this and speeds the **turn off**. To switch current in a coil **on** faster we have to use a higher voltage. We then have to add another resistor to limit the final voltage across the coil and transistor. This circuit is shown in Fig. 9.3c and is the complete coil driving arrangement. We have a bias resistor on the base of the transistor

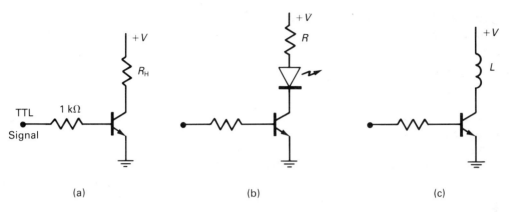

Fig. 9.1 Direct transistor outputs

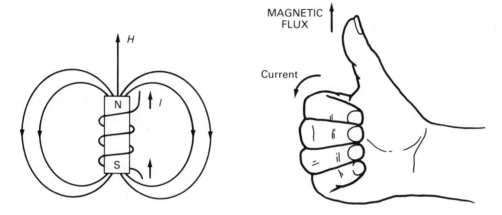

Fig. 9.2 Electromagnetic field formed by coil

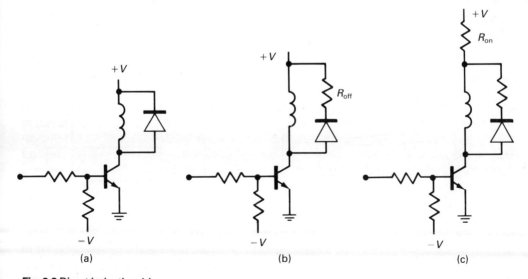

Fig. 9.3 Direct inductive drive

to ensure a clean turn off voltage there, and could add a zener diode in the resistor diode chain to let us cut things closer to the transistor limits. The turn off speed-up resistor is chosen such that:

$$R_{off} \cdot I_{load} + V > V_{bk} \text{ (the breakdown voltage of the transistor).}$$

The single transistor driver is adequate for many purposes, but it is limited by maxima for voltage and current. Figure 9.4a shows a reed relay, housed in a convenient DIL package being driven by a single open collector transistor. A reed relay is made of two thin ferromagnetic strips, with contacts on the ends, housed in a sealed, nitrogen-filled, glass envelope. Application of a magnetic field by proximity of a permanent magnet, or by using a coil, bends the reeds together in about a millisecond to make a contact. The reed relay is small, reliable and significantly faster than ordinary relays.

If we need drive capabilities of more than 30 V or about 50 mA then a single transistor will not suffice. To get more current gain from the milliamp or so a microprocessor gives us, we can employ the Darlington pair arrangement shown in Fig. 9.4b. In packaged form we can easily get half an amp from this; the only disadvantage is the slightly higher input threshold to the circuit. Seven or eight such circuits can be packaged in standard DIL format giving a convenient driver but there will be an overall package power-dissipation restriction.

The development of V-channel metal oxide silicon transistors (VMOS) has brought together the desirable switching characteristics of field effect transistors and high power, high frequency operation. These are now often found as final output power switching transistors.

We will return to switching mains alternating current later, but as well as digital switching analog or linear outputs will be needed. Using the D-A circuits of Chapter 5 will produce the basic value which can then be amplified to the desired voltage and current. There is a small problem, however, one must ensure that the output of a D-A is always correct. With the ladder types a number of bits may have to be changed at the same time, and the associated transistors may not operate at the same speed. This happens particularly if all bits change as shown in Fig. 9.5. If the switches are faster switching to one than switching to zero then with a change from 0111.. to 1000.. we will momentarily pass through 1111.., an incorrect voltage. The solution outlined in Section 5.6 adds a sample/hold after the D-A. The

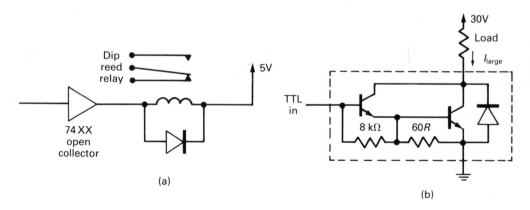

(a)

(b)

Fig. 9.4 Reed relay and Darlington drive

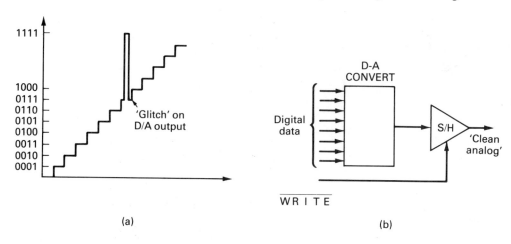

Fig. 9.5 Smoothing D-A outputs

previous output is **held** while the new output is set up. The sample/hold is then set back to sample the output of the D-A and it slews to the new value. Having a clean analog output we use linear amplifiers to get the desired magnitudes. The problem of all linear operation is that the transistors must operate in their **forward active region** and so will dissipate much more power. If we *switch* then our transistor is either cut-off or saturated (fully on). If it is cut off then, though the full voltage is across the transistor, the current is minimal as it is only the leakage current. If the transistor is on then, though the full current flows through it, the voltage across the transistor is very small as it is just the saturated collector–emitter voltage. **Linear operation** has both noticeable current and voltage and so a much larger transistor is needed for the same load.

Thus the pulse-width approach of the stochastic D-A converter has yet another advantage. Assuming that the load is of a type which can be driven by a variable pulse-width rather than a variable voltage then it is simpler and cheaper to employ this technique.

Sometimes it is required that an input signal is recreated as an output. If the input were sampled every 'T' seconds, for example, then an output of a modified value would also happen every 'T' seconds. What form should this modification take? If the input value is simply output directly then data will be output during a very short time (aperture) and then held for a longer time (sample-time). This is called a **zero order hold**. The variables in the equations of Section 3.2 are simply swapped over to calculate the effect which appears as a stepped waveform. It will be a good approximation if the signal changes slowly, but not so good for fast changes as the held value remains until the next output time. If some extra stored 'information' is used then an attempt could be made to predict the way the output should be changing. Extrapolating from two samples gives a prediction of the rate of change (slope) as well as an initial value for the next period. This **first order hold** copes much better with rapid change but not so well with small changes or when the variable changes in direction.

When control of AC power is required then different techniques are used. These are all replacements for older mechanical relay switches. The use of solid-state components in place of the earlier mechanical contactors or relays gives much greater reliability due to the absence of moving parts. Figure 9.6a shows the AC mains input voltage which supplies the load via a suitable control circuit.

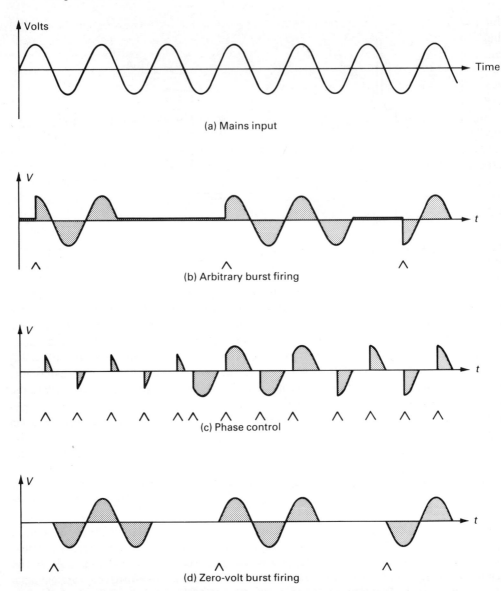

(a) Mains input

(b) Arbitrary burst firing

(c) Phase control

(d) Zero-volt burst firing

Fig. 9.6 Power switching techniques

Simple on/off control of a load is handled through a silicon controlled rectifier or **thyristor**. These are three-terminal devices which exhibit very high impedance when turned off. A high current can flow from the **anode** to the **cathode** if a small current is applied to the **gate**. A gate current of a few tens of milliamps will switch on a main current of many tens of amps. When the gate current is removed the anode to cathode current **continues** to flow. Both currents must be removed for the thyristor to turn off. This makes thyristors suitable for highload AC control but the current is rectified by the thyristor so only half-cycle power gets through to the load. With suitable control a parallel inverse pair of thyristors can give full AC power. A related component, the **triac**, may be thought of as two thyristors back to back but with gates connected. A small current turns the triac on. This is applied to the gate and can be of either polarity. Remove this gate current and the main AC current will then stop. A back-to-back pair of thyristors with separate controls actually gives better surge current capability and high frequency operation, but at the cost of significantly more complex control.

So far the alternating current through the load is only being switched on or off at arbitrary times as in Fig. 9.6b. If this mechanism is used for control purposes there will be considerable hysteresis and also a large amount of electrical interference, caused by switching the current in the load. Finer control is needed and there are two possibilities for part-power controls. These are **phase control** (Fig. 9.6c) and proportional or **burst firing** (Fig. 9.6d). Where the load has a very slow response time, such as in a heater, then a burst of cycles of full amplitude can be fed through it. A period without power connected is then followed by another burst and so on. The ratio of on (burst) to off times determines the power in the load. This technique, also called proportional control, is not suited to higher frequency operation such as AC motor control or light dimming where the harmonics of the process would produce jerky motion or visible flickering.

To operate at higher frequencies, up to the supply frequency itself, we need to switch within a cycle rather than in numbers of cycles. This is termed phase control. For a burst the power would normally go on as the supply passed through the zero point. With phase control this is delayed by a set phase angle until a gate trigger pulse is applied. The thyristor or triac then turns on and the current rapidly increases to catch up with the supply cycle, as is shown in Fig. 9.6. After the main supply passes through zero the thyristor or triac turns off. It will not turn on again until the next delayed gate pulse. The total pulse power through the load is still a percentage of the maximum available power but, whereas burst control uses ratios of whole cycles, phase control gives smooth control at higher frequencies. The main disadvantage is the rapid increase in current as the load current catches up with the supply. These fast-slewing transients cause more electrical interference and noise, particularly radio frequency interference, which has to be restricted by techniques from Chapter 10.

As higher loads are switched when the load voltage is non-zero the problems become progressively more serious. Ultimately, to ensure lowest noise, we must switch only as the power passes through zero and so use burst mode. The burst mechanism can be improved by spreading out the pulses we have uniformly. This is much like the stochastic technique discussed in Section 5.4. Any switching component will perform better if there is no voltage, or only a very small voltage across it when it switches the current into the load. The aim when replacing older parts with solid-state components is to provide the best possible simulation of the mechanical relay with its infinite OFF resistance, zero ON resistance, and excellent isolation between the control signals and the load currents.

The **AC solid-state relay** is just such a unit, being an encapsulated group of components satisfying the requirements above. A wide range is available from many manufacturers with load currents of two, ten and twenty five amps being the most common sizes. Figure 9.7 shows the major parts of a solid-state AC relay. The input signal is an ordinary logic signal compatible with TTL or CMOS logic. A small fraction of the main supply from an isolated attenuator can be taken to a zero-crossing detector. The output from this and the input signal are gated together to give a signal when power is to be connected to the load AND the supply is crossing through zero. An optical isolator links this signal to the control circuit and hence to the thyristor or triac itself. This has a parallel **snubber** circuit (resistor and capacitor) to remove any residual surges. Resistance when turned off is some megohms,

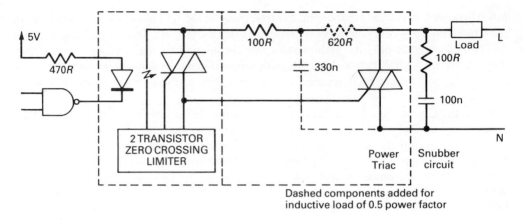

Fig. 9.7 Solid state AC relay

switching speeds are in the millisecond range, and isolation of a couple of kilovolts is achieved. AC solid-state relays are excellent substitutes for their mechanical counterparts.

9.2 Actuators and Motors

Physical movement is provided by actuators and drive motors which convert energy from an electrical or fluid source into a force or torque. To get high efficiency we need high torque and drive stiffness, with low friction and inertia. The effects of friction and inertia on the primary drive are increased in proportion to the square of any gear ratio. If the torque is high to start with then less gearing is needed to increase it. Fluid drives (**pneumatic** or **hydraulic**) are commonly found in industrial applications because they give high force or torque with low inertia. Hydraulic motors and linear rams are used for high power, high performance systems and give excellent acceleration and deceleration. Pneumatic systems do not have such good characteristics but are cheaper. The speed of operation of fluid drives is dependent on the flow of the fluid. This is controlled by varying the size of the orifice of a fluid-flow valve which in turn is controlled by an electrical actuator or motor of some kind. Consequently, though the final drive may be a large hydraulic ram there will have been many intermediate stages in its control. An advantage of fluid drive is that high torque can be achieved without needing high electrical currents at the drive point which could give rise to noise interference in other equipment. Disadvantages are that more maintenance is usually needed to ensure dirt does not block valves or orifices, and a fluid power source is needed. In an industrial environment such pumps may be available anyway.

Linear hydraulic rams are cheap, simple and easy to use for movements of half a metre or so. If too long a travel is attempted the compressibility of the hydraulic oil becomes a noticeable factor. The principle of operation is shown in Fig. 9.8. Rotary motion is generated by forcing the hydraulic fluid past angled valves mounted around an axle.

Linear motion can also be produced by electromagnetic operation only. The device to do this is called a **solenoid** and consists of a stationary coil and a movable iron plunger or armature. When current is passed through the coil an electromagnetic field is set up, as in Fig. 9.2, and this pulls the plunger causing the linear movement or stroke. Direct current gives the best efficiency but, if driven continuously, the coil heats up and the force and stroke drop to half their full range. Hence lower duty cycle (on/off times) are preferable. A typical twelve volt, one amp solenoid generates a movement of only a few centimetres and the force produced would not exceed a kilogram.

By far the most common rotary devices are electrical motors. They come in all sizes from smaller than a cotton reel to bigger than a car. They can be split into four categories:

Alternating Current motors (Synchronous or Induction)
Direct Current motors (Permanent magnet or various windings)
Universal Motors (AC or DC)
Stepper motors

Stepper motors are digitally driven and so are of particular interest for interfacing to microprocessors. These will be described further on. Alternating current motors, particularly synchronous ones, have some good features. Once up to speed a synchronous motor rotates at a known rate very accurately. It has nearly constant torque from start-up

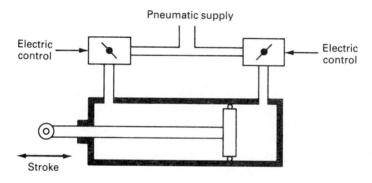

Fig. 9.8 Pneumatic piston

until reaching synchronous speed. A short time after start-up the speed can be relied on to be constant, confirmed by a simple incremental sensor.

Direct current motors can run at higher controlled power, limited only by the ratio of torque to inertia. This limit is set by the armature current causing heating or magnetic saturation. The smallest DC motors are usually permanent magnet types. This type cannot be made very large as the magnet becomes too difficult to construct and use. A motor operates by having a conductor with a current passing through it placed in a magnetic field. The magnet formed by the current in the conductor tries to align with the fixed magnetic field and, if free to move, rotates to achieve this. In an AC motor the current reverses naturally and so the conductor-coil continues to move giving a smooth rotation. In a DC motor a commutator, made of metal segments on the rotor and stationary brushes, can switch the current to reverse it. Of course we could swap everything over. The rotor would be a permanent magnet and the stator would be the coil with the reversing current flowing through it. This current could then be switched externally without needing a mechanical commutator, an arrangement as in Fig. 9.9.

The force on the conductor is the product of the magnetic **field** density times the conductor **length** times the **current** passing through the conductor. So, for greatest force we must increase all three. However, in a DC motor increasing the current increases the torque and increasing the field increases the rotational speed. Thus the way in which the magnetic field is generated both sets the motor characteristics and gives control of the motor.

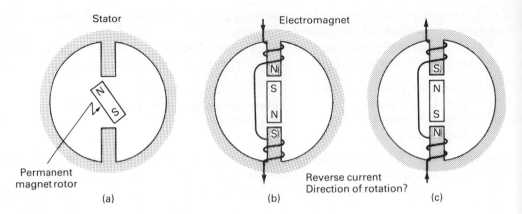

Fig. 9.9 Simple motor principle

If a permanent magnet is used then there is no control over the field so the speed has to be controlled by pulse width or by linearly varying the DC voltage across the windings. The latter gives slightly more even control of these small motors. The field may be generated electromagnetically by a coil. This can be supplied from a separate source from the main armature winding. If, however, the same supply is used then the coil can either be in **series** or parallel, usually called **shunt** wound in this latter case.

If the field coil is in parallel with the armature then it can be made of much thinner wire, have higher impedance, and take much lower current than the armature current. This gives medium power motors for fans, pumps or similar applications. If, instead, the field coil is in series with the armature then it must pass the same current. This means it will be made from a few turns of the same heavy gauge wire and this gives motors with high starting torque. Much greater control over the motor operation can be achieved by using separate supplies, when either the field current or the commutated armature current can be controlled to give the desired effect.

DC motors do not have to have a commutator to reverse the current flow in the armature coil. The necessary alternation can be arranged by electronic means, quite possibly by a microcomputer. These motors are called DC **brushless** motors. They are more reliable than commutated motors as there are no brushes or slip-rings to wear out. Speed control of ordinary non-synchronous motors is commonly arranged by a rotational position sensor providing feedback to alter the motor power source(s). This is more important in brushless motors as the correct phase (coil) must be energized depending on the position of the motor in its rotational cycle. There is a way of providing the feedback without needing a separate transducer. Brushless motors have their coils driven sequentially and each generates a reverse voltage—a back e.m.f. For example, with three coils one will be driven at a time to cause the rotation. The back e.m.f.s of the other two (undriven) coils can be monitored and used as feedback inputs to the microcomputer which is driving the motor. Of course the back e.m.f.s are only present when the motor is running so to start up an arbitrary coil must be energized to get the rotor moving.

The motors most commonly found in daily life are those used in domestic appliances, vacuum cleaners, food processors, power drills etc. These are constructed on the same general lines as serial DC motors but can be driven from AC mains. They are called universal motors.

All of the motors described so far have one big drawback. They can be started, stopped and have their speed controlled, but due to inertia when we stop them we do not know where they will stop. Accurate position control is a great problem if the benefits of rapid movement are to be retained. The brushless motors give a clue. With coils driven separately there seems to be more control over the motor.

9.3 Stepper Motors

A stepper motor effectively converts digital pulses into proportional, rotational movements providing a controlled-position motor. Figure 9.9 shows a permanent magnet motor with two poles. When the current is reversed (by electronic switch) the electromagnet reverses. This forces the permanent magnet to rotate to align to the new magnetic field. The problem of single reversal is that when moving from Fig. 9.9.b to 9.9c we do not know in which

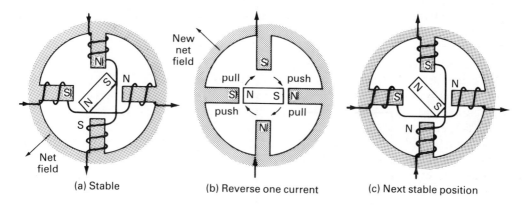

(a) Stable   (b) Reverse one current   (c) Next stable position

Fig. 9.10 Four pole stepper motor principle

direction the magnet rotated. Adding a second pair of coils gives us four stable positions instead of two. In fact, the stepper principle would work with three stable positions (coils and poles) but it is easier to explain a four-phase motor. It is also easier to draw. Figure 9.10a shows the arrangement: with currents flowing down and across, two north and two south pole magnets are formed. Note the direction in which the coils are wound. The net magnetic field points south along the large arrow, consequently the permanent magnet motor aligns its north pole that way. If the current is reversed in the vertical pair of coils then their magnetic poles reverse. The net magnetic field rotates by ninety degrees clockwise and the forces then acting on the rotor are shown in Fig. 9.10b. The rotor is then pushed and pulled to its next stable state aligned with this new net magnetic field. This state is shown in Fig. 9.10c and from here the motor can be stepped on to its next position by reversing the current in the other (horizontal) coil pair. Equally it could be stepped back to the previous position by reversing the original change.

So the principle is very straightforward and the stepping very accurate. Errors are non-cumulative and are usually within 10% of the step angle. A very appealing property of the stepper drive is that if we double the number of poles on the stator we double the number of stable states. The number of connections to the coils from outside the motor is not increased as each new coil is connected in parallel with one of the original ones. Every fourth coil is connected in parallel. When there are more than four poles the explanation of operation is a little more complex than the static argument. When the current is reversed in a group of coils, the magnetic field reduces to zero before increasing to its new value. During this time the remaining poles push the rotor towards the next stable state.

Doubling up poles and connecting their coils in the correct parallel arrangement can be taken a long way. Starting from three poles giving steps of 120 degrees or four poles giving steps of 90 degrees we can get down to steps of 0.36 degrees while keeping to about 10% of step-angle accuracy.

So far the discussion has assumed that the current in the coils will be reversed. This is called **bipolar** drive but needs quite complex switching, at least two transistors at each end of each coil. A simpler arrangement exists. If each coil is wound double (bifilar wound) then one half could be energized to get North/South magnetization. Energizing the other half would give South/North magnetization. Both coils should not be energized at the same time. One end of each half-coil is connected to a common point, normally to earth. Four separate, single-transistor switches are then connected to the other ends of the half-coils. Two are selected to be driven at any time, and the sequence required to step in either direction is the same **gray code** as used in Chapter 2. The full drive circuitry is shown in Fig.

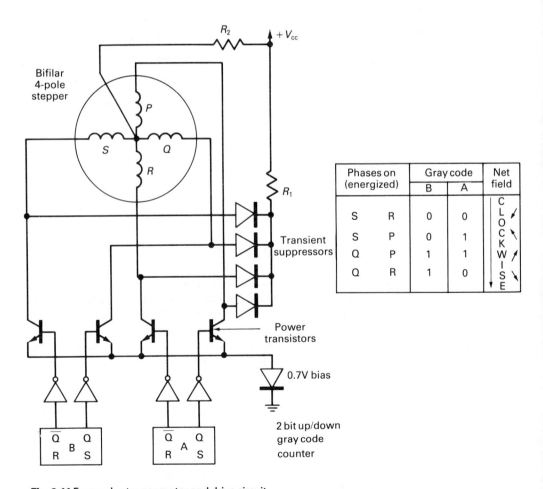

Phases on (energized)		Gray code		Net field
		B	A	
S	R	0	0	C L O C K W I S E
S	P	0	1	
Q	P	1	1	
Q	R	1	0	

Fig. 9.11 Four pole stepper motor and drive circuit

9.11. The table shows the contents of the gray code counter and the coils which are energized by this **unipolar** drive arrangement.

Integrated circuits such as the Philips SAA1027 are available which include all the logic, protection and unipolar drivers to interface small four-phase stepper motors to **step** and **direction** signals. This chip can drive 350 mA per phase and has high noise immunity inputs. With additional drive transistors and protection diodes larger motors are easily handled.

Speed control is easily handled by the rate at which pulses are applied. Certain rates could cause resonance and so must be avoided or alternatively the load inertia and damping must be changed. The range of rates for start-up and for use after the inertia of the load is moving are different. Figure 9.12 shows graphs, typical of those a manufacturer would publish, of pull-in and pull-out rates and torques. The **pull-in rate** is the maximum switching rate which

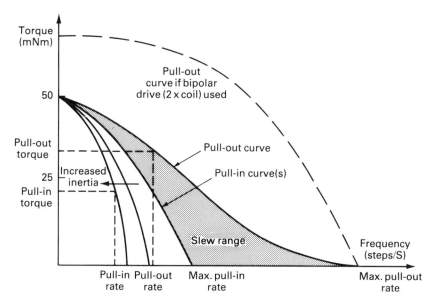

Fig. 9.12 Typical stepper motor characteristics

a frictionally loaded motor can follow without losing steps. The **pull-in torque** is the maximum torque that can be applied to the motor shaft when starting at the pull-in rate. These both change for different load inertias as shown. A permanent magnet stepper always has an attractive force between the rotor and stator poles even when unpowered. This is called **detent** torque and means that the motor cannot be moved from its rest position unless forced by an external torque greater than this. The **pull-out rate** is the maximum switching rate which a frictionally loaded motor can follow without losing steps once it is running at speed. The **pull-out torque** is the maximum torque that can be applied to the rotor shaft when running at the pull-out rate. When a rotor is stationary but the stator coils are energized then the maximum steady torque which can be applied without causing continuous rotation is called the **holding torque**.

Use is made of the fast starting and stopping capability of steppers in many applications. Typical stepping rates, when run at pull-in, are from 200 to perhaps 2000 steps per second. If accelerated and decelerated gently by variable rates then maximum rates of two to four times these can be reached without missing a step. A key benefit of a stepper is that it can run without a position sensor (open loop) and still be accurate. Of course if steps could be missed because of the load inertia then a position sensor is needed. To get the fastest stepping rates a resistor can be put in series with each phase coil and a capacitor in parallel with each phase coil and its associated transistor switch. Such a **compensation** network allows the capacitor to charge up when a phase coil is not driven and then discharge quickly into that coil when it is turned on. This gets the current to build up faster so the motor can step faster. Stepper motor operation can be further optimized by using an optical sensor to ensure pulses are applied as soon as possible. Damping can be controlled by synchronizing stepping pulses with the mechanical position of the motor. The sensor can also be used to adjust rates for high-speed slewing when unloaded, and to give particular patterns of acceleration or deceleration.

There is another type of stepper motor, the variable-reluctance motor. It has a rotor of soft iron with a number of poles which is not the same as the number on the stator. When stepped, usually by a sequence of pulses, the rotor moves to align with the direction of the path of minimum magnetic reluctance. This type of motor can step very fast. There are also hybrids of the two types. So, using the techniques outlined, control of the real world can be accomplished by combinations of movement and switching.

10

Environmental Constraints

This chapter is devoted to a collection of topics which, taken together, determine whether a system will operate reliably in practice and then whether it will be allowed to be run at all! Interaction via environmental factors is a two-way process. Such factors undesirably affect the operation of a system. The operation of a system has side effects which undesirably affect the environment and hence other equipment. So both sides of the problem will be studied to determine the design rules which will limit the unwanted interaction to acceptable levels. A simple sub-division by the location of the sources of problems is most helpful. A problem can originate in one of three places. It could originate in the outside world, for example, noise interference. It could be caused and communicated within a given system, or it could be a very local problem with a particular circuit or connection. The first topic to consider affects all three regions and is noise. Noise is any unwanted signal occurring within a system or emanating from it. When a system is designed it should always be assumed that it will be operated in an electrically noisy environment, and not just in the 'clean' laboratory where it was designed or tested.

10.1 Noise and Shielding

Noise can be **random** or **repetitive**, occurring **continuously** or in **isolated bursts**. It may affect **current** or **voltage** and may occur at any frequency from DC to extremely high frequencies. A particular source may generate noise over a **narrow** or **wide band** of frequencies. The strategy for minimizing the effects of noise depends upon the location of its source. If the noise is generated internally then the designer's aim is to control and limit the mechanisms causing it. If the noise source is external then the aim is to reduce its ingress and subsequent pick-up.

Once noise can be observed in a system it is already a composite signal from many sources so the first step is to isolate and identify each noise source. With knowledge of the mechanisms which produce noise, design rules can be introduced to ensure that systems are both tolerant to received noise and can meet the statutory restrictions on emitted noise. Figure 10.1 shows the major noise source mechanisms and the frequency bands they occupy. The sources are ranked approximately by the ease of solution, those which are easiest to handle being at the top. Noise occurring at frequencies below about 30 MHz will only usually be conducted, whereas from 30 MHz up to the gigahertz regions the noise may be conducted (along metallic or semiconducting paths) or may be radiated through the air. Radiation can occur at lower frequencies (down to 100 kHz or so) if a system contains unshielded, free wires carrying switched signals. This is bad design practice (for other reasons discussed further on) and so should be avoided. Again, different techniques will have to be employed to minimize the original production and to restrict any subsequent travel by the two mechanisms.

There are a number of terms used to describe noise signals. Radio frequency interference

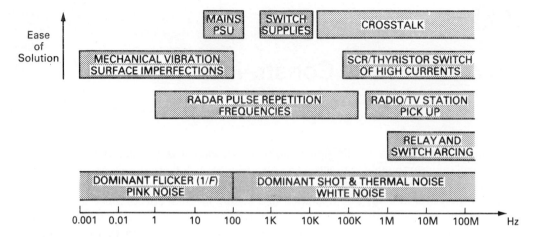

Fig. 10.1 Noise sources

(RFI) has mainly high frequency, low energy components, but they may have significant amplitudes. Electromagnetic interference (EMI) is usually much lower frequency noise but commonly has high energy. Electromagnetic pulse noise (EMP) is a burst signal caused by high power switching or nuclear explosions. Such bursts may contain considerable amounts of energy over a wide frequency spectrum. Electrostatic discharge (ESD) is caused following the accumulation of sufficient electrostatic charge on a surface which has no conducting path to ground. Once the charge has built up enough then an arcing discharge will occur through the path of least resistance, i.e. any unprotected equipment!

In the digital parts of a system there are normally large thresholds between the one and zero states, so the effects of noise are far less noticeable. However, because the rates of change of signals (the rise and fall times) can be so fast the production of noise may be at its worst. When this is combined with the analog sections, having input signals of only a few millivolts and no noise margins, a difficult design challenge is created.

Considering the internal sources first, so-called **white noise** contains the same amount of power in each Hertz of bandwidth. It has Gaussian amplitude distribution, the highest amplitudes having the lowest probabilities, and is not periodic. There are two major mechanisms causing it. **Shot noise**, also called Schottky noise, is a noise current caused by the fact that current flow is not a continuous process. Current flows by the movement of individual electrons which are discrete, charged particles. The root mean square (r.m.s.) value of shot noise over a chosen frequency bandwidth 'f' is given by:

$$I_s = \sqrt{2qI_bf} = 5.64 \cdot 10^{-10}\sqrt{I_bf}$$

where I_b is the bias current flowing and q is the charge on an electron.

Thermal noise, also called Johnson noise, is a noise voltage caused by the random movement of thermally charged carriers in any resistive path. This effect of the second law of thermodynamics is particularly important when it occurs in the resistors used in the inputs to operational amplifiers. Its r.m.s. value over a chosen bandwidth 'f' is given by:

$$V_t = \sqrt{4kTRf} = 1.28 \cdot 10^{-10}\sqrt{Rf} \qquad \text{(at room temperature)}$$

where k is Boltzmann's constant, R is the resistance and T is the absolute termperature.

To minimize the total white noise the bandwidth should be limited to that really needed and source resistances and bias currents should be as small as other considerations will permit.

At lower frequencies, below the region of thermal and shot noise dominance in operational amplifiers, there is a region where the noise increases as the frequency is decreased. This is normally called **flicker** or $1/f$ noise and in low-frequency applications is the most critical internal noise source. Extrapolating the plots of flicker and combined white noise until they meet defines a so-called **corner frequency** which separates the dominant white and pink noise regions. For low noise operation this corner frequency should be as low as possible for the obvious reason that below it the noise increases as $1/f$. The noise content due to flicker is equal in each decade of bandwidth so the total flicker noise over a given band is:

$$V_n(f) = K\sqrt{\ln(f_h/f_1)}$$

where K is the noise content in a decade which has negligible white noise e.g. 0.1 Hz to 1 Hz. Typical corner frequencies are about 200 Hz for the ordinary 741 type operational amplifier and as low as 7 Hz for very low noise versions such as the S725. There are other internal noise sources commonly caused by imperfections in the semiconductor production process or materials. For example, a low frequency burst bias-current change is produced by faults in the surface of processed wafers. This is normally called 'popcorn' noise.

In all cases the bandwidths of the analog sections of a system should be limited to the band actually required to ensure no additional extraneous noise is included with the signal. The low voltage analog parts of a system can be badly affected by the internal noise sources if the rules mentioned are ignored, but both analog and digital parts of any system are subjected to noise generated outside individual circuits. They are also affected by noise generated outside the entire system and again it is beneficial to identify the sources to determine the appropriate corrective action.

This latter group of noise sources, completely external to a system, are almost without exception the artefacts of another electrical system. A high current load which is switched on or off causes transients in its power supply lines due to their inductance and/or capacitance. Efficient switch-mode power supplies have become very popular with the makers, and users, of mini- and microcomputers. They generate much less heat than equivalent linear power supplies but can generate significant noise between 1 kHz and 10 kHz. Fluorescent lighting systems produce noise at 100 Hz or 120 Hz depending on the local mains frequency. The spectrum of switching supplies shows a discrete set of pulses at the fundamental switching frequency or harmonics of it. By contrast phase control, using thyristors and similar devices which were discussed in Chapter 9, shows a continuous interference spectrum which is a bit harder to filter out as described in Section 10.3. Any high current flowing in a line parallel to a signal line will produce noise on it even if it is some distance away. Over only a metre of a parallel run of lines a metre apart, a millivolt of noise will be caused from the current supplying a 3 kW load.

Anything which causes a spark or arcing will radiate an electromagnetic wave as noise. All the earliest experiments in 'wireless' transmission used exactly this principle to signal with! Relay and switch arcing and motors with commutators or slip-rings fall into this category. A similar type of noise pattern is produced by radio and television stations which deliberately 'pollute' the ether. An ordinary audio hi-fi system often experiences break-through of police, fire or taxi transmissions and similar effects can be measured in unprotected digital systems.

Radiated noise enters a system as some form of electromagnetic wave. Any varying electric field generates a magnetic field and any varying magnetic field generates an electric field. Far away from their source the ratio of the two fields (E/H), which is called the **wave impedance**, is 377 ohms. Where the wave impedance significantly differs from this value then one or other mechanism is dominant. This happens closer than about 1/6th of a wavelength from the source and, not surprisingly, is called a 'near-field' effect. Where the ratio is higher than 377 ohms then it is the electric field and hence capacitive coupling which is dominant. Where the ratio is less than 377 ohms then it is the magnetic field and hence inductive coupling which is dominant.

Pretty well all internal radiated-noise coupling in high-speed microprocessor-based systems is near-field as $W/6$ is five metres at 10 MHz, and still half a metre at 100 MHz. Capacitive coupling (E-field) generation is from parallel or straight conductors. To minimize this coupling, free wires and close-running parallel signal lines should be avoided. Additionally a (preferred) alternative route can be provided for the noise by an earthed shield put between the source and the victim of the pickup as shown in Fig. 10.2. This is called a **Faraday shield** and it restricts the E-field propagation by reflecting it. A shield which fully encloses components is called a Faraday cage. This type of shield does increase the capacitance to ground and so somewhat slows the rise and fall times of switched signals. Reflection occurs by re-radiation of the field and is a surface effect so is largely independent of the thickness of the shield.

The reflection loss is approximately equal to log $(\sigma/F\mu)$dB where F is the signal frequency, and σ is the conductivity and μ the permeability of the shielding material.

With very low noise operational amplifiers having such tiny input currents even the leakage **across** the printed card surface has to be taken into account. Screening must enclose the entire circuit and all of its external components including feedback etc. The layout of connections and components becomes yet more important and a special printed-circuit track layout called a **guard** is used. This is a track which surrounds the input pins and is connected to the shield (and ground). The track itself forms a shield and prevents any stray currents crossing the surface to the inputs.

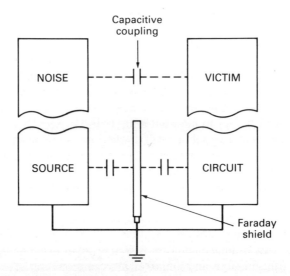

Fig. 10.2 Faraday shield

Inductive coupling (*H*-field) generation is from loop geometries (forming coils) and the amount is determined in part by the area enclosed by the loop. To minimize the pick-up layouts which form looped connections must be avoided, or if they cannot be avoided then their enclosed area must be minimized. For example, a full ground plane means that no ground loop area is formed. A 'grid' power supply layout minimizes the area of the small loops compared with direct supply and ground wiring which commonly forms quite large loops. This is discussed further in Section 10.3. To reduce inductive coupling a shield is again inserted between the source and the victim circuit but in this case the shield operates by absorbing the magnetic field. The currents induced in the shield as well as causing reflection by re-radiation also experience loss due to the resistance of the shield material to the currents.

The loss of power $(P = I^2 R)$ weakens the noise field and the absorption loss is approximately

$$t\sqrt{\sigma\mu F}\ \text{dB} \qquad \text{where } t \text{ is the thickness of the shield.}$$

Probably the most potentially damaging form of noise for modern semiconductor devices is **electrostatic discharge** (ESD). It is less of a problem in the UK due to the maritime, and hence humid, climate. It becomes a serious problem in those areas with a continental climate, or in buildings with air conditioning, and much dryer air. A build-up of electrostatic charge is possible on almost any material especially nylons, acrylics and silks. The human body, with a capacitance of about two hundred picofarads and capable of being 'safely' charged up to tens of kilovolts, is frequently the culprit. Remember that a four kilovolt potential can produce a visible spark. However, it is normally the rate of change which makes ESD so damaging. The current pulse from an electrostatic discharge can rise at **four amps per nanosecond** and, from a body charged to 30 kV, can reach 50 amps for a duration of perhaps 30 nanoseconds.

The final set of noise sources are external to a particular group of circuits under consideration, but internal to the system. Power-supply-induced problems are discussed in Section 10.3, so the major problem here is the pickup of one signal on another path. **Crosstalk** is mostly by *E*-field coupling so the Faraday screen approach is used to reduce it but, as before, the screen efficiency falls as the source signal frequency increases. Though the layout of ordinary printed-circuit boards may exhibit the problem, crosstalk becomes acute when many signal lines must run in parallel over a significant distance as, for example, in a backplane bus. The main factors affecting the amount of crosstalk are: the signal rise times, the path length and the geometry and termination of the path. From the arguments above it was shown that capacitive coupling increased as the signal frequency increased, and thus as the rise times shortened.

Crosstalk can be measured by driving a signal across two conductors and measuring the voltage on a free, undriven line. This can be done at the source end, giving a 'near-end' figure which is almost directly proportional to the driven peak voltage. It can also be done to give a 'far-end' figure and this appears as a spike of opposite polarity to the driven signal which increases as the path length increases and as the rise time shortens. Such a crosstalk pulse has a duration twice the rise time of the driven signal.

The Faraday screen approach can be applied to printed circuit tracks and to cable interconnections. With a signal having a rise time of 2 nanoseconds, 0.3 metre of adjacent wires of ordinary flat-ribbon cable produce a far-end crosstalk of 12% of the driven signal. This can be reduced to 10% by including alternate ground wires and to 6% by using twisted pair cable. If a complete earth plane was incorporated in the cable the far-end crosstalk would drop to 3%. Just as significant a decrease could be produced, however, by increasing the rise time of the signals to between 10 and 12 nanoseconds. This would reduce the crosstalk because the source frequency would be lower. To give the best protection against radiated noise a totally enclosing shield is required. There are obviously design problems in meeting the 'totally' because inputs, outputs and power must connect somehow. If the continuity of the screen is broken then the flow of current is impeded and both re-radiation and absorption are reduced. Where a box must have a lid, or a cabinet must have a door, continuity can be retained by using a conductive gasket to give (nearly) continuous contact. Metallic knitted mesh and metal-doped sponge rubber are two of the materials used for this. If cable access must be provided then it is far better to drill a row of holes separated by their own diameter than to cut a long slot which has the same total hole-area. It is easy to see why. Just draw a set of parallel lines to represent the current flow and see how much they would have to bend to get out past the widest part of the long, thin slot, and how little they would be impeded passing between the holes. Joints and seams can also cause problems. If they do not have continuous contact, i.e. they are bolted or spot-welded, then a similar rule applies as to slots and holes. A given length of slot (or diameter of hole) will only give protection against waves of length less than a twentieth of this. Thus if F MHz is the highest noise frequency against which protection is sought then $L_{max} = 15/F$ metres for commercial applications. More stringent criteria are applied to military applications where $L_{max} = 6/F$ metres is used.

Conductive coatings are a convenient, cost-effective solution to the problems of shielding for plastic enclosures. Acrylic resin-based paints containing nickel loading are now commonly used as they are cheaper than silver-based and more stable than copper-based paints. They are applied with ordinary spray equipment and give good conductivity and attenuation. Typically a thickness of 0.025 to 0.05 mm will give 40 dB absorption of RFI between 5 MHz and 2 GHz due to the permeability of the nickel. It also gives a surface resistivity of less than 1.5 ohms per square, compared to 0.01 for silver dope and 0.5 ohms per square for a copper dope.

Once an ESD surge has eluded or penetrated the screening and has got onto a signal line then protection against actual component damage can be arranged by connecting a zener diode from the line to ground. Connecting two zeners 'back to back' in series (i.e. in opposite polarity) gives protection against surges of either sign. A zener diode has the normal diode characteristic when forward biased, that is it conducts and as the forward voltage increases the current through the diode increases. In the reverse direction only a tiny leakage current flows, but in a zener when a given voltage is reached an avalanche effect causes the zener to pass whatever current may flow while remaining with the constant voltage across it. It is this voltage-limiting effect which restricts the voltage from an ESD or EMP to a safe amount by diverting the excess current to ground. Commercial devices are available which exhibit the effects of the two series zeners but are constructed as a single component to make them respond quicker.

Having considered how to defend a system from noise one must also ensure that its emissions are acceptable. **Statutory restrictions** are placed on the emission of electro-magnetic interference from electronic equipment in many countries. The standards usually taken for design are those promulgated by the Federal Communications Commission (FCC) in the USA and the Verband Deutsche Electrotechniker (VDE) which is the society of West German Electrical Engineers. The relevant specifications are FCC 47 CFR part 15J (1983) and VDE 0871/6.78 (amended 1982). They are shown in simplified form in Fig. 10.3. The

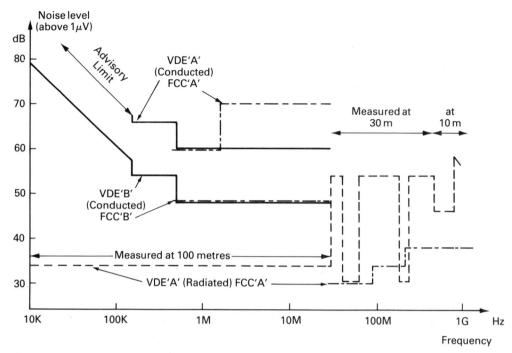

Fig. 10.3 Simplified electrical emission requirements

limits are defined for conducted EMI at frequencies from 10 kHz to 30 MHz and for radiated RFI at frequencies to above 1 GHz. They are also specified for various categories of apparatus and of particular interest for interfacing are: class A covering industrial and business systems and class B covering domestic electronic equipment. The VDE limits are released by the Deutsche Electrotechnische Kommission as DIN standards and are rather more strict than the equivalent FCC codes for the most part, so designing for one should ensure compliance with the other. It is a pleasant coincidence that reducing pickup using the rules described also reduces the production of noise. As with attempts to conform to any standard, the latest version of the standard itself must be the reference.

10.2 Interconnection of Circuits

It is relatively easy to understand how a particular (integrated) circuit will operate, but what happens when two or more circuits are connected together? There are a number of factors to consider and they can be summarized by:

- The mechanism(s) to drive a signal onto the connection path.
- The way the signal is transferred and the properties of the interconnection path.
- The interaction between the received signal and the receiving circuit(s) at the end(s) of the interconnection.
- The effect of the loading (input current drain) of the receiving circuit(s).
- The interaction between unconnected but adjacent circuits or interconnection paths.
- The interaction between unconnected circuits which share the same power supply.

A large proportion of the problems in system design come from the interconnection of circuits to form sub-systems and their interconnection to form complete systems. Almost any typical schematic diagram of a system could be interconnected in such a way that it will be **un**-reliable yet meet all the requirements of the schematic. This is a fundamental limitation of the notation and the shorthand used. For example, most schematics do not show how the power supplies should be interconnected. Arrow-heads pointing up and 'ground' symbols serve to hide this part which is absolutely fundamental to the reliable operation of a system.

All of the problems are met at printed-circuit board level with existing systems but, as systems get faster, the problem is simply moved inside the integrated circuits as well. That, however, is another story and some good texts on IC design are covered in the bibliography.

In an **ideal** case an output switches from one state to the other. A signal propagates at the speed of light (c) to the input circuits connected to the output. There, the input voltage transits from the original level to pass the threshold to the new level. If the circuit is combinatorial then the output remains holding this new level stable until the next change is required. If the circuit is a bistable then the output only has to hold the level stable for the minimum set/reset time. This must be more than two gate delay times for an ordinary bistable. Of course, this description assumes perfection, so consider what really happens.

Firstly the output will not switch in zero time. This is because of the physical size of the transistors and the parasitic capacitance caused by their proximity to other components and connections.

Secondly the path itself has characteristics which make driving it a non-trivial exercise. A small aside is to note that each joint in the path as well as adding series resistance may create a thermally generated e.m.f. in an identical fashion to the thermocouples in Section 2.5. A copper-to-solder connection has a 3 microvolts per degree C thermal e.m.f., and while this is small in the digital sections it may be quite noticeable in analog signals.

Thirdly when a signal reaches the intended receiver(s) it will have some loading effect, and it may come as quite a surprise to realize what can happen.

Finally, the time at which a pair (or more) of signals reach a receiving circuit may have very unexpected results.

Consider a pulse being driven onto a path, then its instantaneous voltage and current could (in theory) be measured at any point along its path. If the characteristics of the path are constant then the relationship between current and voltage will also be constant and be given by Ohm's law:

$$Z = V/I$$

where in this general case Z is a complex number, the size of the impedance, rather than the usual pure resistance 'R'. If V and I are the instantaneous measurements for a given cable then Z is normally written as Z_o and called the **characteristic impedance** of the cable. Different cable configurations yield different characteristic impedances and different increases to the loop area of the connection. In general it can be shown that Z_o is approximately equal to $\sqrt{L/C}$ and the delay per unit length of line is approximately $\sqrt{L.C}$ where L is the inductance and C the capacitance of the cable. Figure 10.4 shows **coaxial** cable, **twisted-pair** cable and printed-circuit board **strip-line** connections. Coaxial cable is made with a central conductor separated from a braided outer screen by one of a number of insulating dielectrics. A coaxial cable should carry the same current in the outer and inner conductors so the fields produced by them should cancel exactly. The cable adds no extra

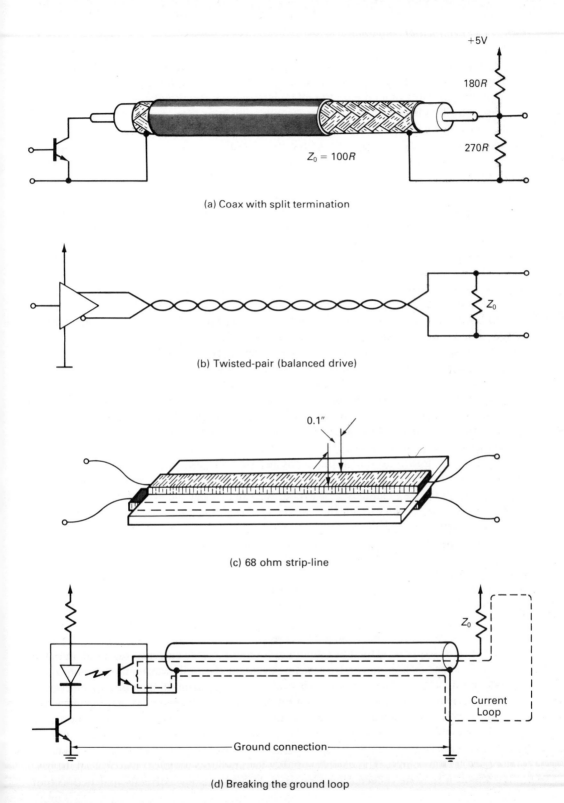

(a) Coax with split termination

$Z_0 = 100R$

+5V

180R

270R

(b) Twisted-pair (balanced drive)

Z_0

(c) 68 ohm strip-line

0.1"

(d) Breaking the ground loop

Z_0

Current Loop

Ground connection

Fig. 10.4 Coaxial, twisted-pair and strip-line connections

area to any loop connected through it so long as the currents are actually equal. If, however, both ends of the shield are earthed then an extra path exists. Low frequency signals may pass through this common ground path but all high frequency signals will only propagate in the coaxial cable. This is because it presents the minimum impedance path to these frequencies as it encloses a smaller area and so has a significantly smaller inductance. Typical coaxial cables have characteristic impedances from 50 to 100 ohms with 75 ohms being the common cable used for television aerials etc. One way of removing the low-frequency path caused by the double earthing is shown in Fig. 10.4d using an optical coupler as described in Section 4.1.

A cheaper cable is made by twisting two insulated conductors together. The characteristic impedance of twisted-pair cables varies from 50 to 150 ohms and is determined by the thickness and properties of the insulation and the tightness of the twist. Running the connectors side by side (rather than concentrically) must add a small amount to the loop area of any connection. This unfortunately gives greater impedance, greater RFI from the cable and greater susceptibility to external interference. A twisted-pair cable can be driven in the balanced way described in Section 7.1 to reduce common-mode noise.

Strip-lines are formed whenever a copper track on a printed circuit board runs parallel to a ground line. For glass-epoxy board of 1/16th inch thickness a 0.1 inch wide track on one side and a full ground plane on the other give a characteristic impedance of 50 ohms. If the ground plane is reduced to a ground track of the same width (0.1 inch) as the signal track then the impedance increases to 68 ohms. Running two similar parallel tracks 0.04 inches wide on the same side of the board with a separation of 0.04 inches increases the characteristic impedance to 100 ohms. As any of these strip-lines forms a rectangle there is a significant increase in the loop area of the connection, hence lower bandwidth and more noise susceptible operation ensues.

Now, back to assuming that a pulse has been driven onto a path (any type) with a characteristic impedance of Z_o. All will be well as long as there is no change in $Z_o(=V/I)$. But what happens if Z_o changes as it would, for example, at the end of the cable. If the impedance of the receiving circuit (the load) is equal to Z_o then the same voltage and current still satisfy Ohm's law and all the power from the pulse is dissipated in the load. In fact the pulse cannot tell the difference between such a load and a semi-infinite length of cable of similar construction to that used in the path. A pulse starting just before the load and travelling off to infinity (to the right or the left, but not both!) would never be seen again so that is what is (correctly) expected of this **matched termination**. If the load impedance does not match the characteristic impedance of the cable then there will be too much (or too little) power for the load.

The two extreme cases where the load is zero (a short circuit) or infinite (an open circuit) show what must happen. If the load is infinite (ohms) then the current through it must be zero, yet 'I' amps (at 'V' volts) arrive at the open circuit. The only way to make 'I' into zero is to add 'minus I' to it. This is exactly what happens, the current in the pulse reverses and flows back down the line. This is called a **reflection**. In this case the voltage remains the same, the current is reversed and the pulse 'reflects' back off the incorrect termination and travels back to the source. A rough analogy to this process is a long taut string. If it is plucked quickly the pulse (bump) in the string can be seen to travel to the end and then 'bounce' or reflect back to the source. The electrical pulse, having been reflected, travels back to the source where it may be reflected back again if the source is not correctly matched to the path.

If the load is zero ohms then there can be no voltage across it, yet 'V' volts arrive at the short circuit. The only way to make 'V' into zero is to add 'minus V' to it. Consequently the reflection reverses in voltage as well as in direction and an inverted pulse travels back to the source.

In general, with a line having characteristic impedance Z_0 and terminated by a load of Z_t then:

$$\frac{V_{(reflect)}}{V_{(original)}} = \frac{Z_t - Z_0}{Z_t + Z_0} \quad \text{the \textbf{voltage} reflection coefficient} \quad \text{and}$$

$$\frac{I_{(reflect)}}{I_{(original)}} = \frac{Z_0 - Z_t}{Z_0 + Z_t} \quad \text{the \textbf{current} reflection coefficient.}$$

This mismatch effectively limits the data rate available on a given path as a new pulse cannot be driven onto the path until the reflections from the previous data pulse have died away. If the original pulse has reflected and has not been reduced sufficiently by the multiplicative effect of successive reflections and the finite resistance of the path then it could be interpreted as another (new) data pulse if one were expected then. So that is what happens at the receiver termination, but what happens at the source? If the source is not correctly matched to the transmission path then firstly, optimum power transfer between the driver and the path does not take place and secondly, there will be the additional reflections off the source back to the receiver.

The solution to the problem is to terminate all transmission paths correctly with the matching impedance. This is quite difficult for most types of logic circuit in common use as will be shown below, but there is a small loophole available.

Electromagnetic waves propagate at $1/\sqrt{\epsilon\mu}$ metres per second where ϵ the permittivity and μ the permeability are characteristics of the material of the path. Light travels in a vacuum at an approximate velocity of $c = 3.10^8$ metres per second. Most signal paths such as those discussed have permeabilities very close to that of a vacuum. Their permittivity, however, may be 3 to 4 times that of a vacuum so their signal propagation velocity is only 50–60% of c.

If the transmission time between the driver and the receiver is small compared to the rise time of the pulses being transmitted then the reflections can die away without upsetting the signal. Some simple calculations using the reflection coefficients show that for quite badly matched lines a single reflection back and forward along the path is sufficient to let the signal reach its 90% level which is adequate for the vast majority of digital inputs. In these cases the restriction on the transmission time being small compared to the rise time can be relaxed to it being half the rise time or less. Remember that at the speeds discussed a signal travels a metre in five nanoseconds so the distances for which the loophole applies are quite short for fast rise time signals. For example with a 1 ns rise time the limit is 10 cm for the 'relaxed' rule and only 2–3 cm for the worst case. It should be noted that the majority of backplane bus systems in small computers do not have drivers which match the very low impedance loads and so are not capable of sourcing the current needed immediately. They rely on reflections to build up the necessary voltage on the bus lines to give correct states.

A further serious design restriction due to interconnection is that output circuits cannot supply arbitrarily large currents to input circuits and still remain within the designed voltage threshold limits. Hence the loading imposed by one circuit on another and the capability of a circuit to drive a load require consideration. Looking at a driver considered earlier (Fig. 7.11a) and simplified in Fig. 10.5 (i.e. a single output being driven correctly) then the output circuit either has its top transistor **or** its bottom transistor turned ON. If the top transistor is ON then the power supply acts as a **source** of current to pull any inputs up towards the power supply voltage. If it successfully pulls the inputs above the 'high' threshold then it is driving correctly. This action is called **sourcing** and the current which can be sourced is limited by the chain of components from the power supply to the output wire. Conversely, if the output has its lower transistor ON then the lower transistor draws current from the inputs of connected circuits and acts as a **sink** to ground (the negative end of the power supply in this case). This action is called **sinking**, and the current which can be sinked (sunk?) is limited by the size of the lower transistor. If the transistor cannot handle the current then too great a voltage will be developed across it and the output voltage will exceed the permitted threshold for 'low' levels. It should be noted that the sourcing and sinking capabilities of output circuits are not symmetric for many types of logic.

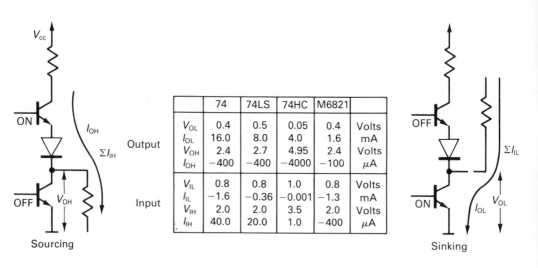

	74	74LS	74HC	M6821	
V_{OL}	0.4	0.5	0.05	0.4	Volts
I_{OL}	16.0	8.0	4.0	1.6	mA
V_{OH}	2.4	2.7	4.95	2.4	Volts
I_{OH}	−400	−400	−4000	−100	µA
V_{IL}	0.8	0.8	1.0	0.8	Volts
I_{IL}	−1.6	−0.36	−0.001	−1.3	mA
V_{IH}	2.0	2.0	3.5	2.0	Volts
I_{IH}	40.0	20.0	1.0	−400	µA

Fig. 10.5 Source and sink loading

A simple calculation from the tables is all that is required to ensure that a given output is capable of driving the loads put upon it. Referring back to the problems of matching terminations we saw that typical connection paths had characteristic impedances of 50 to perhaps 200 ohms. Now at five volts this means a current of from 30 to 100 milliamps is needed. Clearly, from the table with Fig. 10.5 the outputs do not provide this much. The circuits must either be driven slowly enough for the 'rise-time versus transit-time' argument to apply or they must be used to drive to a lower voltage. This can be done by using a potential divider of two resistors to set a lower voltage to which the line is driven. Another way, as was seen in Section 8.4, is that high speed bus interfaces can now use special low-voltage drivers (BTL) and receivers to ensure correct matching is practical. At the chip level Emitter Coupled Logic (ECL), which is the fastest off-the-shelf logic at present, also has a much smaller voltage swing and so is much easier to match to its transmission paths.

And so we come to the problem of the relative arrival times of signals at the intended receiver inputs. Because of the finite propagation delays and the increase in rise and fall times due to input capacitance, two signals may arrive at different times from those expected. The problem can be made worse by having one of the signals come from the outside world and having the other be an internal 'clock' signal. There is now no phase or timing relationship between them. They are truly asynchronous. Transients can be caused on a signal line due to noise, but in this case transients can be caused by the 'correct' operation of the circuit.

Consider an S-R bistable, such as that shown in Fig. 2.3a, with the set and reset signals true i.e. $S = R = 0$ volts, a switch from both sides to ground. If the **set** is released first (S becomes 5 volts) by more than two gate delays then the circuit will end up reset. The opposite will happen if the **reset** is released first. But what happens if they are both released (set false, to 5 volts) within two gate delays of each other? The circuit then does not have time to set (or to reset) before the other signal is removed and so may do neither. A simple mechanical analogy is an inverted pendulum (a pencil hinged at its point). It can be pushed over from one side to the other, but it is possible for it to stand up vertically, balanced in unstable equilibrium. In theory it could stay there for an indefinite time. Luckily, the slightest puff of wind (equivalent to some electrical noise) will eventually tip it back to one of its stable states. Any of the 'glitch' patterns shown in Fig. 14.9 could be caused by this process.

To see a little more deeply into the problem consider a set of traffic lights. The rule is simple: green light for go, red light for stop. The problem is the inertia, or braking distance of a car, which is analogous to the set/reset time of a bistable. A car is approaching a green light when the light turns red. The driver has to decide if the braking distance is such as to allow a safe stop before the junction or if he or she must accelerate across the junction to the other 'state'. The solution appears to be to set a point before the junction at which to decide. But this simply moves the problem back, because now at the point when the driver checks red/green for stop/go the light changes as he or she is checking and a decision must be taken as to which colour the light is. This extra decision takes some time, by which time it is too late to make the original decision safely. The driver brakes and ends up stationary in the middle of the junction. Adding an amber light moves the problem back further still, but, by the same argument, does not solve it. There is a finite probability that the time taken to make the decision will be long enough to leave the output state uncertain. But we have now seen how to minimize (not solve!) the problem. Firstly the tiny gust of wind must be made very small, or the system be made very unstable in the 'meta-stable' state so that it cannot last long. Secondly more time must be allowed to make the decision.

Two mechanisms are employed to reduce the asynchronous 'glitching' problem. One is to increase the loop gain of the bistable so that a slight change on an input is quickly amplified to a large change in output state. The second is to use two bistables clocked a small time apart so that should the first go 'meta-stable' it will have returned to a stable state before the second is clocked. This gives a far higher probability of a glitch-free output from the second bistable. This is not a certainty and so all truly asynchronous systems have a mean time before failure, even if it is ten years or more.

10.3 Power Supply to Circuits

The mains electricity supply is often taken for granted, particularly in some countries where it is a very reliable source. It is not totally clean though and contains spikes, surges and RFI. **Spikes** or transients are the voltage deviations caused when electrical machinery and power station generators are switched. They can reach 2500 volts but last for only a few milli-seconds, yet a typical office could expect a thousand spikes in a year. **Surges** are much slower deviations but may change the nominal mains voltage by 20% for as much as half a second. Negative surges or **sags** are more common than positive deviations. It is believed that three-quarters of computer malfunctions are attributable to power supply and power line disturbances.

Basic protection is achieved by a mains filter which reduces spikes, transients and RFI by 60 to 65 dB. They are available in a range of sizes from 3–20 amps. The simplest filter puts a capacitor in parallel with the supply to act as a low-pass filter. Adding an inductor (choke) in series with the supply gives a further damping effect by absorbing high-frequency energy in magnetization of the core. The use of toroidal-wound inductors with balanced windings connected in both live and neutral lines produces opposing fluxes in the core. This prevents saturation and gives improved attenuation of symmetrical noise. Figure 10.6 shows a typical mains filter design with additional capacitors to earth to cope with asymmetric noise on the live or neutral line only.

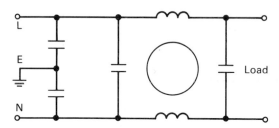

Fig. 10.6 Simple mains filter

Protection is improved by using a ferro-resonant isolating transformer. These remove sags and surges if the supply frequency is constant, as in the UK, but are not suitable in areas where noticeable frequency fluctuations occur. The transformer has a tuned resonant secondary winding which produces a current-limited, regulated sinusoidal voltage output. These circuits, sometimes called constant voltage mains regulators (CVR), also reduce spikes, transients and RFI and can hold the supply up for about half a cycle in the event of a very short interruption. They operate with efficiencies from 70 to 85%.

Reserve power (charge) is available in most systems' power supplies to cope with very short interruptions and still keep the low voltage parts, including any microprocessor, operating. However, an interruption to the mains supply of only 1–2 milliseconds can cause a thyristor or silicon-controlled rectifier to switch thus giving incorrect phase control.

The ultimate in protection of the main supply is given by uninterruptable power supplies (UPS) which are available, at a cost, in a range of sizes from 100 VA to 10 kVA and above. An UPS removes all fluctuations and provides battery-generated back-up against drop-outs and short-term power cuts from 10 minutes to some hours depending on the unit. The UPS has a rectifier connected to a battery charger and on to a battery. This is then connected to an AC inverter circuit to regenerate the correct sinusoidal 'mains' output with total efficiencies of 80–85%. System power may be continuously drawn from the inverter and the battery. An alternative arrangement is to have a switch so that under normal circumstances the direct-mains supplies the system (albeit with reduced protection) and the battery and inverter are only switched in when power failure is detected. The charger is cheaper as it never has to supply both charging and system supplies and larger loads can be supplied particularly if some load can be shed on power failure. Having the switch arrangement can also allow the normal mains to act as a back-up for a continuous UPS in the event of the UPS failing!

Power supplies are high-power devices when compared to the rest of a microcomputer and are either linear or switch-mode regulators and so are quite complex to design. For data logging and other small portable units battery supply may be used with advantage. The purpose of a power supply is to provide absolutely stable supply voltage(s) regardless of the input mains voltage and frequency, the output load and environmental factors such as temperature. The design of power supplies falls outside the scope of this book and for small numbers it will always be cost-effective to 'buy in' a supply if there are any complex specification requirements. There is an argument as to whether the final regulation stage should be distributed (to be local to each board) or be centrally located. The advantages of distributed regulation are that each regulator handles less power, is simpler to design, localizes failure and distributes heat generation over the whole system. The major disadvantage is that each regulator may regulate to a slightly different voltage and so logic levels might not be the same. This causes problems if an input voltage is actually higher than the supply to a circuit. Also, if a fault occurs it may not be detected unless there is a complex monitoring circuit which loses the simplicity. Cooling has to be arranged over the whole system and some area is lost on every board. With modern, switch-mode supplies, centralized supply is more accurate and reliable at present.

The distribution of power around a system or board is very important as a badly laid-out supply will cause intermittent 'faults' in the powered circuits. There are some important considerations for distributing the power from the supply to the circuits:

- Voltages are measured **between** two points, not **at** one point.
- Two points connected together (by wire) are not at the same potential if a current flows between them.
- Connecting two points by more than one path creates an inductive coupling loop.
- No path can pass an infinite current, the limit is set by the resistance of the path.
- A current flowing in a path cannot change instantaneously, charge cannot be moved any distance in zero time.

A single point may be designated as the signal ground reference point. It can be the ground at the output of the power supply for example. If a single line is run from this point to a number of circuits (a series ground) then there will be different 'ground' voltages for each circuit. This may not matter in the digital parts of a system as noise thresholds are quite large. Also this 'ground' potential of each circuit will vary depending on what the other circuits are doing. If a circuit in the middle of the line takes more current then the 'ground' potential of one at the far end of the line will go up. Of course an equivalent drop is experienced on the voltage supply rail. Using a series connection for analog, digital and high power switched loads will guarantee that the system does not work. Separate ground lines of sufficiently low resistance are needed to each circuit with different noise characteristics and current requirements. On a printed-circuit card then ideally every transmission path would be a constant characteristic strip-line. This can only be arranged by having a (nearly) continuous ground plane. This allows the earth currents to run under the signal tracks associated with them and also has the benefit of providing nearly constant earth potential over the surface. The power rail(s) could also be distributed this way with advantage, but there is a significant increase in cost. Hence this would only be done with very high speed circuits.

If a continuous ground plane cannot be provided, for example if only a double-sided board is to be used, then the gridded-ground approach most nearly meets the requirements. A full grid of ground tracks is laid out, using both sides of the board where necessary. This

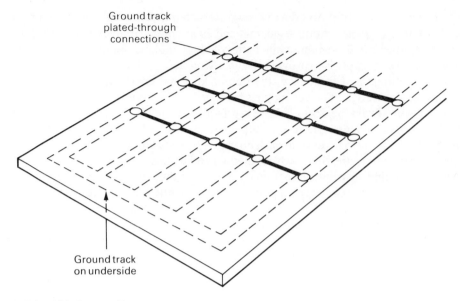

Ground track
plated-through
connections

Ground track
on underside

Fig. 10.7 A gridded-ground layout

minimizes the loop areas and allows the earth currents to take the best path they can. Figure 10.7 shows a gridded-ground layout using both sides of the board.

A basic mains-driven power supply has a 50 or 60 Hz alternating current as its source. A simple microprocessor accesses its store at a rate of 1 MHz and so the supply cannot keep pace with the changes unless there is an extra charge storage mechanism. An analogy is filling a bucket with water as fast as possible. This is the equivalent of the microprocessor suddenly needing charge supplied when it switches all its address lines (e.g. from $7FFF_{hex}$ to 8000_{hex}). If the bucket is filled from a tap turned fully on it will take some seconds (five or more) to fill. The problem is that the supply is a distance from the need and there is a restriction (resistance) between them. If we now try to fill two buckets at once the problem gets very serious as, in five seconds, both buckets can only be half full or one may have grabbed all the water leaving the other starved (thirsty)! Thus in an ordinary microcomputer the microprocessor, store, I/O and ancillary circuits all operate at the same time and all have needs for charge from the power supply which must be met at the same time. One solution is to give each bucket its own pipe and tap. The buckets no longer interact, but they are still not filled fast enough. The source must be placed closer to the need! If we keep a bath topped up to its overflow line then buckets can be randomly filled from it in less than half a second. The tap will restore the level during slack demand periods.

This is the principle of a **Decoupling capacitor**. It is a charge store placed as near as possible to the circuit it supplies. High frequency demands by the IC are met from the decoupler which is kept topped up by current supplied from other, larger, charge-storing capacitors spaced out all the way back along the path to the power supply. The question of how large the decoupling capacitors should be has often got mixed up with where they should be located. Remember that the purpose of these capacitors is **not** to filter out noise spikes put onto the power supply rails by the switching current demands of integrated circuits. The purpose of a decoupler is to ensure that no demand for charge supply by one IC is ever detectable by another. Consequently, decouplers should be as large as is reasonably

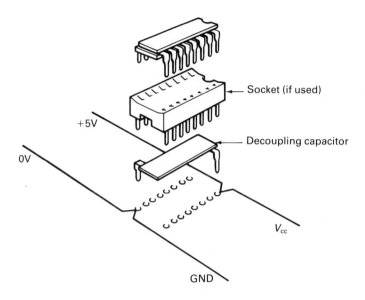

Socket (if used)

+5V

Decoupling capacitor

0V

V_{cc}

GND

Fig. 10.8 Ideal positioning of decoupler

practical. Of course, if they are too large then they will take a very long time to charge up to working levels, rather like filling a public swimming pool from a single tap. Simply work out the maximum charge requirement for worst case switching against the maximum allowable voltage supply tolerance of the chip to get the necessary size. Values from 1 to 10 microfarads are adequate for common microprocessors and their major support circuits.

Many circuits do work, however, with seemingly inadequate decoupling. The reason is that if one circuit is switching an adjacent one is probably not. Thus both circuits contribute to the charge supply for the one, and both have their voltage supply drop close to the tolerance limit. This is not good design practice.

Decoupling capacitors should be located in such a way as to minimize the inductance and resistance of the connection path as these only serve to **reduce** the frequency response, and hence effectiveness, of the decoupled supply. Figure 10.8 shows the ideal location for a decoupling capacitor which adds no area to the loop connecting it to the IC. Side benefits are that less tracking is needed and board area is saved. Decouplers of this form are now commercially available.

It should now be apparent from this chapter that for straightforward designs simply following all the design rules will yield reliable, working end products. For systems which operate at high speeds, with low power, or approach extreme limits in any other way, complex calculations or design iterations are needed to give a satisfactory result.

11

Microprocessor Input-Output

Microprocessors exhibit many advantages over traditional machines, but there are some drawbacks. Size and low cost are the obvious advantages, but the ease of handling input–output must be one of the most important. Traditional mainframe machines had expensive central stores and processors. I/O devices were slow by comparison and so considerable effort was put into delegating the I/O transfer load to subsidiary processors in a hierarchic arrangement. These units called **channels** were interposed between the I/O devices and main program operation. These channel units needed their own programming to run efficiently.

Thus, complex sequences of instructions had to be arranged and passed through to the channel and adaptor hardware. This I/O organization was found to be suitable for only a few specific devices. Many different types of channels (selector, multiplexer, block multiplex, etc.) were produced to attempt to resolve the problem but only served to make it yet more complex.

The more modern approach is to put the minimum of hardware between the actual I/O device and the (micro)processor. The devices regularly cost more than the entire processor and store. Hardware is only added between the device and processor where it is necessary to keep up the speed of transfer for very fast devices. The input or output operations can then be programmed in the same flexible manner as any other part of a program, thus fitting the user's needs better, and at reduced cost.

Hence it is unfortunate that all high-level languages designed in the past (particularly those currently common on micros) have had their input–output functions tacked on to the language in an entirely ad hoc fashion. Programmers are thus unable to arrange their

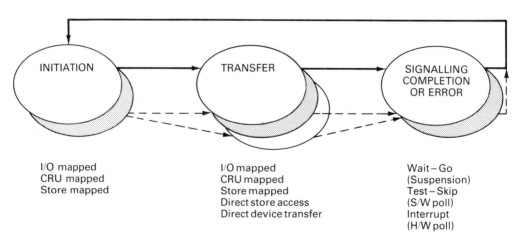

INITIATION	TRANSFER	SIGNALLING COMPLETION OR ERROR
I/O mapped	I/O mapped	Wait – Go
CRU mapped	CRU mapped	(Suspension)
Store mapped	Store mapped	Test – Skip
	Direct store access	(S/W poll)
	Direct device transfer	Interrupt
		(H/W poll)

Fig. 11.1 Microprocessor input-output phases

program I/O in a clean, logical and provably correct manner. The situation is changing, however, and the omissions (such as any I/O at all in ALGOL-60) are being rectified in newer languages. Modula-2 is one of the first to address the problem seriously and attempt a solution. However, it is from a study of the basic operations of input–output that a complete solution for the future will come.

It can be observed that input operations should have no side effects i.e. they can be fully 'functional'. Conversely, output operations will commonly have side effects as they affect the real world and the effect cannot necessarily be undone. If a system is given the same program with the same set of inputs a second time, without any intervening outputs, it will compute the same results. As a result, more care is needed on the output side to ensure that the effect achieved is the effect desired.

Any I/O operation consists of the same three phases: **initiation, transfer** and **signalling completion** as in Fig. 11.1. It is in the way these phases are organized that the simplicity of microprocessor I/O is found.

11.1 Initiation of Input–Output

Any I/O transfer must be started so this phase is essential. The other phases can be formed into a single operation as, for example, an output to a digital-to-analog converter could be achieved by a single instruction. It is not possible to make initiation automatic and there are only a few methods by which the programmer, via his program, can cause the necessary action.

In general a machine instruction takes the form:

[Operator] [Source Operand(s)] [Destination Operand(s)] [Next Instruction]

with various parts implicit because of the chosen machine architecture.

Thus there are two basic methods of initiating I/O.

  (i) By employing specialized **operator(s).**
  (ii) By employing specialized **operand(s).**

These give rise to the two basic forms of initiation, respectively:

  (i) **I/O mapped** input-output.
  (ii) **STORE mapped** input-output.

The traditional, and perhaps obvious, form is **I/O mapped** and uses specialized instructions to cause input or output initiation (or transfer or testing). The operand(s) of the I/O instructions specify the device or 'port' address and the source or destination of data. Figure 11.2 shows the layout of locations and some typical instructions used internally and for I/O. It is obvious that there are two address spaces, one the ordinary store and the other, usually much smaller, for I/O. The microprocessor's interface reflects this. The data bus and low order address lines are common. The high order address bits are exclusively used for the store. It is the control lines which split the spaces, there being separate **Read/Write** and **I/O Read–I/O Write** signal lines. There is normally a 1:1 mapping of device registers to ports (locations) in the I/O address space, i.e. there is a unique address for each device register to be accessed.

Turning to the instruction set. For I/O mapped I/O there are only a very few instructions, usually just **IN, OUT**, but perhaps also **TEST**. The first and simplest form, for example from

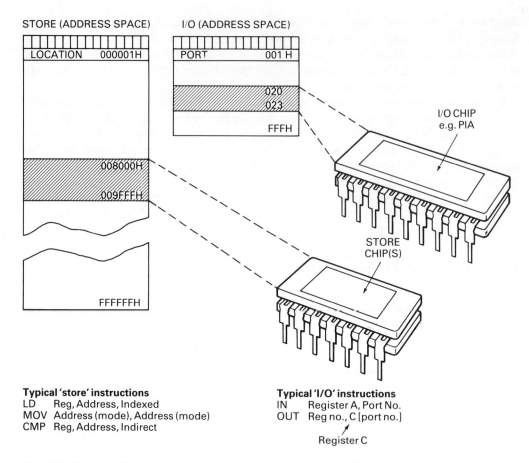

STORE (ADDRESS SPACE)

LOCATION   000001H

008000H

009FFFH

FFFFFFH

I/O (ADDRESS SPACE)

PORT   001 H

020
023

FFFH

I/O CHIP
e.g. PIA

STORE
CHIP(S)

Typical 'store' instructions
LD   Reg, Address, Indexed
MOV   Address (mode), Address (mode)
CMP   Reg, Address, Indirect

Typical 'I/O' instructions
IN   Register A, Port No.
OUT   Reg no., C [port no.]
                    ↑
              Register C

Fig. 11.2 I/O mapped input-output

the Zilog Z80, operates between a fixed register (**A**) and a **port** number fixed in the instruction at compile time. This is inadequate because normal run-time alteration is not possible. Hence the operations could not be handled by re-entrant procedures for a group of similar devices. The second form shown allows for a choice of registers and for the port address to be held in a given register (say **C**). This is more usable by an operating system but requires the registers to be pre-loaded so that simple output to a D-A can take three instructions.

I/O mapping collectively has the advantages that it is easy to analyse and trace, and that it is unlikely that transfers would occur accidentally (by conjuring up instructions!). There are also a number of disadvantages:

- instruction sequences are needed for simple atomic operations
- extra instruction codes are required (e.g. In, Out, Test)
- it is inflexible on the 'device' side—transfers are in units of bytes or words fixed by the microprocessor not the device
- it is inflexible on the 'store' side—the full range of address modification is not available, there is a limited range of operations.

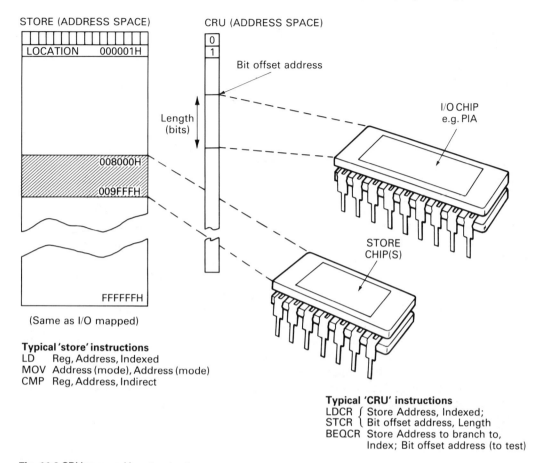

STORE (ADDRESS SPACE)

LOCATION   000001H

008000H

009FFFH

FFFFFFH

(Same as I/O mapped)

CRU (ADDRESS SPACE)

0
1

Bit offset address

Length
(bits)

I/O CHIP
e.g. PIA

STORE
CHIP(S)

Typical 'store' instructions
LD   Reg, Address, Indexed
MOV   Address (mode), Address (mode)
CMP   Reg, Address, Indirect

Typical 'CRU' instructions
LDCR ⎰ Store Address, Indexed;
STCR ⎱ Bit offset address, Length
BEQCR Store Address to branch to,
           Index; Bit offset address (to test)

Fig. 11.3 CRU mapped input-output

It would be possible to correct the last one by not restricting I/O operations to such a limited set of normal instruction operations. No machine has done this yet however. There is an approach to give flexibility on the device side however by restructuring the I/O address space. Most machines work on bytes or words though the better micro architectures have more types, as explained in Chapter 13. Devices work in bits or arbitrary groups of bits. Consider for example the D-As of Chapter 5, or the stepper motor drives of Chapter 9. It seems obvious that the device side should be bit-organized, and there is a system like this. It is shown in Fig. 11.3 and is called **Communication Register Unit** I/O or CRU-mapped I/O.

The CRU considers the I/O address space as a bit-wide register of considerable length. It is bit-addressable on any arbitrary bit boundary and in units of any arbitrary number of bits from 1 to say 16 or 32 depending on the micro. The instructions take the form shown giving the extra device flexibility but usually adding extra addressing modes on the 'store' side, for example using store locations rather than registers as source/destination. A further refinement is automatically to add a register content containing a bit base to the offset in the instruction to give the CRU address of the first bit of the field at run-time.

We can have one last benefit from the CRU approach. As the I/O space is bit-addressable it makes sense to include a TEST instruction, perhaps of the form 'Branch to location XX if the CRU bit specified is equal to zero' (or not equal, etc.). Again automatic base/offset addressing can be used to locate the CRU bit. The interface signals are also interesting. Rather more address bits are used as the CRU is bit-addressable but only a single data line in or out is needed for the synchronous transfers. The arguments in favour of serial rather than parallel transfer have been presented in an earlier chapter but again here the bit seems the natural unit. The full generality of this approach would be achieved if the store side also addressed to arbitrary bit boundaries but these have never been implemented together.

The CRU approach has not been well understood by most designers and so is not very widely used though one range of micros is based on it. Perhaps the reason for its limited acceptance is the availability of another solution which overcomes all the disadvantages in one go.

STORE mapped I/O started to be used in the late 1960s when it was widely realized that the operands could convey all the information required for the I/O operation without needing any specialized instructions. In fact it is a positive benefit that any instruction could be used as an I/O instruction allowing, for example, the direct comparison of an input value with a stored variable or even another input value in one instruction. This is much more comprehensive than just a bit test. Figure 11.4 shows how this method provides a single uniform address space for all operations. This also implies that the full range of data types is available for input-output as for ordinary store accesses. This will greatly simplify the operations for the less common store data types, for example **bits** or arbitrary **bit-fields** which are the most common for 'real-world' interfaces.

I/O devices are simply **mapped** into a part of the uniform address space, their registers appearing to the machine as store locations to be read from or written to. An arbitrary area of the address space is usually set aside for the I/O registers. There is one serious problem though. It is very likely that an incorrect operand may be generated accidentally and so an I/O operation (possibly unsafe) would be executed incorrectly. As an example of how this could happen, consider how, on powerup, personal computers determine how much store is available and then self test it. They simply try all available locations. Similarly problems can occur if run-time array bound checking is not employed and a program addresses outside its planned areas.

To get over this the chips which are designed to be store mapped device interfaces can contain added protection to ensure that the 'accidental' write does not actually perform a damaging I/O transfer. Such protection would normally require two instructions per operation though and so a speed loss would occur. A better and more general solution lies in the architecture of the machine including a correct and convenient store protection mechanism. In the same way that any program (or operating system) store space is protected from any other, so too is any store mapped I/O space. However, as is usual in computing, protection and generality are gained at some loss in speed as the effective operand address(es) of each instruction executed are checked by the hardware.

Thus far it has been implied that a store mapped I/O device register appears identical to any other store location. If *complete address decoding* is employed this is true, but there is an alternative which is common in small systems. If the address is not completely decoded then a number of addresses will actually access the same physical location (or I/O device register). This is called partial or *degenerate addressing*. The reduced decoding saves money but it can be put to use as well. A single location appears to be at a number of (contiguous) addresses. Thus a block move instruction to read from those continuous addresses will actually give the contents of the one I/O location as a sequence of values at times separated

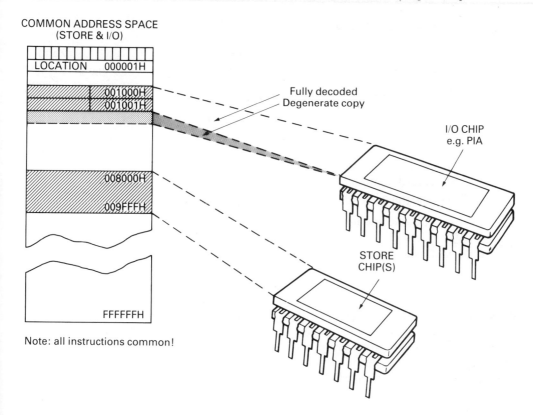

COMMON ADDRESS SPACE
(STORE & I/O)

LOCATION   000001H

001000H
001001H

Fully decoded
Degenerate copy

I/O CHIP
e.g. PIA

008000H

009FFFH

STORE
CHIP(S)

FFFFFFH

Note: all instructions common!

Fig. 11.4 Store mapped input-output

only by the store cycle time. This is a very fast and simple way of reading a sequential burst of inputs from an ADC for example. The samples are taken in at sub-instruction speed without additional hardware. Of course as the degenerately addressed I/O device (or store location) appears at many locations care must be taken to avoid accidental operations on the *degenerate copies* as well as on the 'expected' address.

When using I/O operations in a high-level language it may not be possible to know in advance which style of initiation operations will be used. Under these circumstances the high-level language should have specialized operators (or functions) such as **read** or **write** to initiate I/O. The reason is that while it is easy to map specialized operators into specialized operands, the reverse is very difficult. It verges on the impossible if the operand addresses are calculated at run-time as every store access would have to be checked to see if it was directed to an I/O location before mapping that operation into one handled by an I/O instruction code.

To recap: input/output operations can be initiated by one of three techniques of which **store mapping** is the most flexible and general. Arbitrary areas of store are designated for I/O operations and a store protection scheme is operated by the processor. Any ordinary instruction can be used to perform any I/O operation. The full range of data types, including **bit** and arbitrary length **bit-field** are used to give I/O flexibility. The full range of address modes are used to allow the user to organize I/O initiation in the most convenient manner.

11.2 Transfer of I/O Data

In some cases the act of initiating the transfer will actually carry out the data transfer and no testing for completion is necessary either. An example is the output from a digital to analog converter where the transfer can be done in one write cycle. An eight-bit micro directly interfaced to an eight-bit D-A will achieve this. In most cases though transfers are longer and may need repetitive action to transfer all the data. There are five arrangements for transferring data to and from I/O devices.

- **I/O mapped input/output**
- **CRU mapped input/output**   } **Programmed I/O as in Section 11.1**
- **STORE mapped input/output**

- **Direct Store Access input/output (DSA or DMA)**   } **Autonomous I/O**
- **Direct Device Transfer (DDT)**

The first three can obviously continue to be used to transfer subsequent data for large transfers. Up to this point only the fundamental operations of **read** or **write** have been considered. To support **semaphore** operations we need the equivalent of an **atomic** (indivisible) operation 'read then write'. It is convenient to permit a modification of data between the read and write as well. To allow in-place updating of linked lists without semaphore protection a further stage of writing would be needed in the atomic operation. Finally block moves of data as a linked series of reads or writes are needed. The transfers can interleave with other processing by using the techniques described in Section 11.3. They are limited in maximum data rate as they rely on instruction execution for each transfer. Many I/O devices are slow however and so the instruction execution rate is adequate. No extra hardware is used. At higher data rates these methods both run out of steam and swamp the processor. Extra hardware is then required.

A **direct store access** device is somewhat similar to a traditional 'channel' but now the chip is low cost it is simplified as it only ever needs to handle one device. It effectively provides the block moves of data mentioned above and the major parts are shown in Fig. 11.5. It uses one of the previous three methods for setup to load the **store address**, the **count** of items to transfer and the **control** mode. Once the setup is complete the DSA acts as an autonomous device competing with the processor for the bus lines when it needs to do a store access. Thus it has two control interfaces, one to the store to read or write data onto the bus and the second to do the reverse operation to the peripheral device chip. Additional bits can be added in the mode to permit a selector DSA to choose one of a few I/O devices for this transfer. The bus access mechanisms are discussed in Chapter 8 but commonly the DSA device 'steals' a bus cycle from normal processing for each transfer. Once the entire block has been transferred the DSA completes its operation with an indication that more programmed I/O is now required. There are alternative methods by which the DSA gains access to the store. It may suspend or even stop the processor to gain its access. A better way is to 'steal' store cycles or stretch the clock pulses to fit in its accesses. A trick which gives no degradation is to use some part of the instruction cycle when the DSA knows the processor cannot be accessing the store to fit in its own transfers.

Direct Store Access has a longer initiation time (setup) than programmed I/O but a much reduced overhead per byte transferred. Only a few store cycles are lost rather than many instruction times per transfer. It is well suited to block transfers and as the chips are cheap

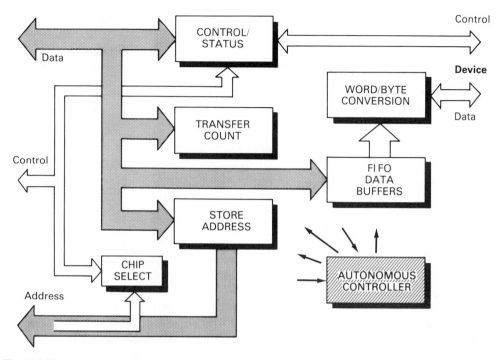

Fig. 11.5 Direct store access

and simple to use may be employed equally well for low-speed as well as high-speed devices. The only drawback is that some DSA controller chips are designed with unnecessary extra complexity and being pin limited have to adopt 'dirty' loading techniques. These rely on set sequences which **must** not be interrupted and overlapping addresses etc. which are awkward.

The final option is a **Direct Device Transfer** or DDT. If store mapping has been employed then a direct store access device is simply given the 'store' address of the other device, and so transfers directly between the devices rather than directly to the store. An additional controller would be necessary if I/O mapping were used to address devices. If we wish to transfer data from one device to another there is no requirement to go via the store as in all previous methods. A DSA transfer can be set up from the device to another and then allowed to run autonomously, divorced from both processor and store. Only bus cycles are stolen when DDT transfers are done. As most processors have some internal cycles when the store is not being accessed, and the bus speed is often fast enough, the DDT time can be completely overlapped and so there is no loss of performance. DSA and DDT controllers can be arranged to sequence one block address/count/mode after another automatically to give continuous operation if needed. Quite complex I/O coprocessors are now being made which provide DSA and DDT functions. In addition they permit more complex mask and comparison tests for completion, and on-the-fly code translation.

11.3 Signalling Completion or Error

I/O devices are sometimes slow by comparison with the store. Programmed block transfers are sometimes made to/from I/O devices. DSA devices complete their operations at arbitrary times. Unexpected errors may occur in a device when a transfer is under way. All of these require the normal sequence of processor instructions being carried out for some other reason to be deviated to cope with the I/O. The signal is asynchronous with respect to the processor's instruction sequence and there are three ways to synchronize the two.

- **Wait/Go**   —a suspended wait
- **Software polling**   —a busy wait
- **Hardware polling**   —a non-busy wait (interrupt mechanisms)

The **Wait/Go** mechanism was found in many simple microprocessors but has lost favour. When an I/O operation is initiated the micro is suspended from normal operation, entering a wait or ideal state until a signal is received on an input pin to reactivate it. The disadvantage of this is just that the processor is inactive and unable to control any other devices. The advantages are that synchronization is automatically achieved, that the system is simple and that it is easy to understand.

Software Polling provides more flexibility but at the cost of complexity as usual! Instead of suspending the processor, it is allowed to continue running some useful computation. This computation is, however, regularly interleaved with programmed I/O to test for completion or error in transfers which have been previously initiated. Multiple transfers can be run in parallel but the difficulty of programming all the tests, interleaved with computation, is obvious. The most common **monitor** program for eight-bit microprocessors (CP/M) is entirely programmed to use software polling, so it can be done. A monitor is a very rudimentary operating system and, programmed in this way, can never respond to any unexpected input and so is useless for any real-time control or logging work.

Analysing the actions performed by the monitor to service all the I/O leads to the conclusion that there are two types of program involved. The application programs initiate I/O but are unconcerned with its detail. I/O handlers on the other hand simply take the individual parts of I/O and link them together to form the complete transfers needed by the applications programs. If the two types were programmed separately and run separately then the 'monitor' would be much cleaner and more powerful. As is often the case, more hardware is the solution. The problem with software polling is the difficulty of programming interleaved computation and I/O tests. The solution is to let the hardware test all the I/O at every instruction.

Hardware Polling is the result, commonly called an interrupt mechanism. At the start of each instruction the interrupt line(s) are tested and if any have been set since the last check, the interrupt mechanism is activated. There is a sequence of actions to be taken to handle this signal and the choices will be examined. On interrupt the processor must:

- **Disable other interrupts (at this priority level)**
- # **Save the current state of the machine**
- # **Establish the cause of the interrupt**
- **Set up the new machine state for handling interrupt**
- # **Enter the new state**

Following execution of the code to handle these interrupt actions the processor must be returned to its main program. The necessary actions for the return are:

- **Leave the handler state**
- # **Restore the saved (previous) state of the machine**
- **Enable interrupts (at this priority level, but hold off!)**
- **Enter the old state (where it was interrupted)**

The '#' marks those points in these sequences which usually give rise to delay and require more detailed analysis. There are four ways of saving the current state of the machine. The state must include the **program counter** and the **processor status** (PSW) but will also include any registers used by the interrupt handling software, and may include all registers if as a result of rescheduling a different process is run after the interrupt service is completed. It is implicit that separate areas of store will be used for the handlers' variables and so all variables of the main program are automatically saved.

11.3.1 Saving the current machine state

We can save the machine state in **fixed store locations**, by switching to an **additional set of registers**, by using a **stack** or by having an infinite set of 'registers' to switch to. This is called **workspace switching** and effectively provides a distributed switchable stack.

The problem with fixed store locations is that an interrupt cannot be allowed to interrupt an interrupt unless some additional saving mechanism is used. Fixed store saving was common on minicomputers such as the classic Digital Equipment PDP-8 and a few early micros but is far too restrictive to be used in current machines.

Switching to a secondary set of registers is better because it is faster to switch than to store, but the same problem remains. The second interrupt needs a different (additional) mechanism. An example of this approach is the Zilog Z80. It has an implementation of a secondary switched register set, but only for some of the registers. This restriction is unfortunately one of a number of faults in its input/output structures.

There are really only two viable arrangements: **storing on a stack** or **switching workspaces.** Figure 11.6 shows the result of normal stacking. It would be possible to have the stack at a **fixed** location with just an offset held in a register. This is inconvenient and provides no benefits as it does not permit the stack to be moved. A full address is held in the **stack pointer** register and this is initialized to point to the first location of the stack. No matter how they are drawn on paper, it is an arbitrary choice in the machine architecture for stacks to operate by increasing or decreasing store addresses. There is only a slight difference in the choice of incrementing (or decrementing) before or after the store operation. It is marginally better to make stacking faster than unstacking where the choice exists. Once an interrupt has been detected those registers which are to be saved are stored in sequential locations starting at the address given by the SP and stepping its content between each store operation. The number of registers saved, above those which must be saved, is again a trade-off. More makes interrupt handling slower. Fewer registers saved means that the programmer must be careful of which registers may be safely used. Some micros e.g. the Motorola M6809 provide both extremes, **full** or **fast**, for the user to select.

A stack is such a useful arrangement for general storage management of last in first out (LIFO) data that it seems obvious to use it here too. There is a better solution though. The reason stacks were the first choice was because **registers** were fast compared to **store** accesses and so needed to be saved on interrupt. Nowadays, as the access times of store chips have fallen so that they approach those of registers, there is no longer a compelling need for separate general-purpose registers. The whole argument for having multiple levels of storage **with different types of access** falls apart. **All** store locations could be treated as 'registers' and all operations would work on store locations. You will recall that all the store variables were saved by switching to a separate area for variables needed by the interrupt routines. The same technique can be used for the 'registers' if they exist in the store as a

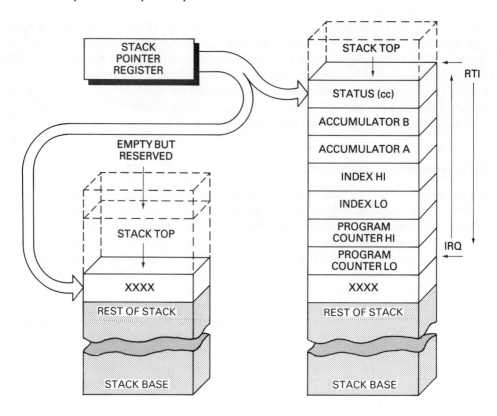

Fig. 11.6 Stack operation

workspace and are all accessed via a single workspace pointer (WSP), as is shown in Fig. 11.7.

The machine only has three registers (PC, PSW and WSP) to be saved on interrupt. The first two are saved in the space allocated in the current workspace. The old workspace pointer is stored in the new workspace (so that the old workspace can be recalled, of course) and the three new values are loaded into the registers to complete the switch. This saves the complete state of the machine in the shortest time i.e. both **full** and **fast.**

The workspace switching architecture can be viewed as either a nearly infinite set of switchable registers or as a distributed stack. If viewed as the latter then it is a very fast stack as it does not require the delay for storing usually inherent in stacking. The only drawback of a workspace-based machine is that every 'register' access is now a store access after the addition of an offset (the register number) to the workspace pointer content (a base). This will be slower for ordinary accesses than a simple register system but does allow a uniform base/limit protection system for all accesses. Thus, although most modern architectures employ stacking to save the curent state, workspace switching is both faster and neater.

11.3.2 Establishing the cause

Interrupt mechanisms operate by testing an input signal or signals at the start of each instruction. There will commonly be many I/O devices causing interrupts and fewer lines to

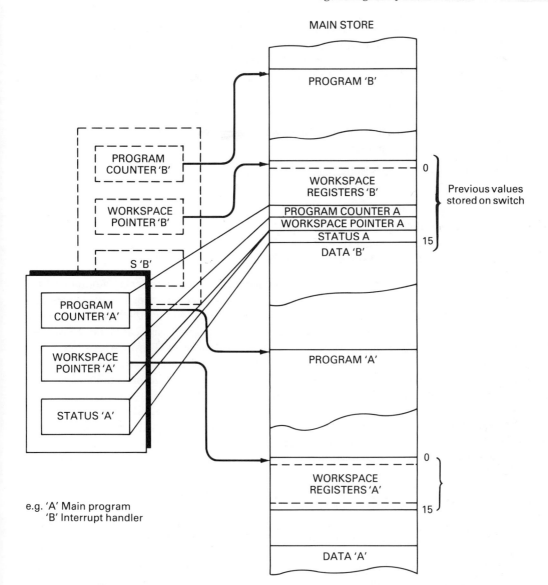

Fig. 11.7 Workspace switching concept

signal them. Thus the machine will know an interrupt has occurred but still have to determine its cause. The system will also have to cope with cases when more than one interrupting device signals within one instruction cycle.

One particular interrupt will require special care. This is the **Reset** or power-on interrupt. It starts the whole machine running correctly and its handler has to set up code for all the others. A complementary interrupt for power failure is very desirable but is not found on most micros. It is important with on-line and real-time control systems to ensure that they fail safely, or hand over to a standby system. The variables and states should also be saved in non-volatile store so that on a subsequent powerup safe running can be restored.

There are also many internal interrupts, usually called **traps** or **exceptions** to indicate store protection violations, virtual store page faults, division by zero, attempts to execute illegal instructions, and many other causes requiring software to handle them.

For I/O devices there are three main ways to arrange the initial incoming interrupt signal(s):

- **single interrupt line**
- **single interrupt line with return acknowledge (daisy chain)**
- **multiple interrupt lines.**

There are then four ways subsequently to determine the cause from the set of possible causes:

- **no further action required**
- **software poll**
- **additional unencoded vector**
- **additional encoded vector (priority encoded).**

Obviously the different schemes give different performance as differing amounts of the chore are handled in hardware or software. With only a single interrupt line all devices have the same chance to interrupt and so there is no priority exclusion handled by the hardware. The addition of a return acknowledge line allows the external physical layout to provide exclusion of less important interrupts by more important ones. The interrupt line is still common (usually driven by open collector) but the return line is passed in daisy chain fashion through each device. The most important device is first in the chain and so on in order. Once an interrupt is signalled the return is asserted and the first device in the chain accepts, but does not forward, the acknowledgement. It is then the only device to respond. The processor cannot adjust this operation except by disabling a particular device. A more flexible arrangement is to have multiple interrupt lines. The physical arrangement spreads devices across the lines to permit the processor to select the priority group(s) it wishes to service.

Once the processor has received the interrupt from the priority level(s) it wishes to service it will have to determine which device caused it. If there are multiple interrupt lines and no more than this number of devices then no further action is required. If the number of devices causing interrupts is greater than the number of interrupt lines then a software poll can be used.

The processor executes instructions to examine a bit or bits in each device to see if it caused the interrupt. Polling is obviously slow, taking many instruction times to find the cause but this is acceptable if the system is not heavily loaded. If we must find out quicker then extra hardware must be added.

The first way would be to use an interrupt acknowledge line to order the devices to signal on the data or address lines which ones were requesting interrupt service. One line is simply allocated to each device without any encoding. This is identical to the parallel poll mechanism of the IEEE 488 interface described in Chapter 8. It is simple but the processor still has the job of sorting out the 'vector' to get to the right service routine. Of course there cannot be more interrupting devices than there are data or address lines. The device could be given the responsibility of providing the complete vector, but this is unreliable and chaos ensues if two devices respond or a device provides an incorrect or corrupted address.

The full solution is to add still more hardware so that on interrupt the highest priority, enabled, interrupting device indicates in such a way that its 'vector' can be used to get straight to the service routine. When the interrupt is acknowledged the device **could** put an address onto the address lines. Whilst this is what we want it is so fraught with unreliability that we must find a safer way to get this effect. As an aside various early micros did this or, even worse, had the device put an instruction on the data bus. We cannot rely on a large number of devices doing this safely so their separate interrupt lines are fed to a priority encoding logic. This provides a single interrupt and the encoded vector of the highest priority interrupting device. The vector is then used as an offset down a table of addresses held in store. The correct address is put into the program counter as the start of the handling routine. This mechanism is shown in Fig. 11.8 and as well as determining the cause also enters the new state. There are other ways of doing this though.

11.3.3 Entering the new state

Many ways have been used over the years to enter the new state. The simplest, now only found on obsolete micros, was to switch the program counter to a fixed value—say location zero. The handler routine (or a jump to it) must be located at that point.

Most systems nowadays use read-only store to hold the basic input-output functions, particularly the bootstrap code to start everything off. The absolute location of the interrupt handler must then be in either in ROM or in RAM. Either way it just won't do! The initial handler needs to be read-only store but the user must be able to produce his own modified version of a handler to be located in read-write store.

An easy improvement is to have the fixed location contain the **address** of the handler. It is still a bit tortuous as the bootstrap code must set the contents of the fixed locations (in read-write store) to the addresses of all the default handlers in the initial read-only store. This begs the question of how the bootstrap is entered in the first place. There are usually many of the in-store vectors, one for each I/O interrupt line, reset and each exception or trap. The obvious places to locate them are either at the top or bottom of the store address space. Having decided that both read-only and read-write store are needed, and that the reset vector should be in read-only store and all the other vectors (the vector table) in read-write store, the author created two solutions. The first solution to this problem was to site the reset vector at the top of store and the other vectors at the bottom, or vice versa. This fits in well with the arrangement of store chips, stacks, and heap usage and store mapped I/O. It is also the fastest solution but has not appeared on micros yet. The second solution was a little slower but has now appeared on commercial micros. Another level of indirection is added by having a 'vector table base register'.

This register initially contains zero so the reset vector is at location zero with the other vectors at increasing locations. The initial read-only store holds these. When, after the reset has entered this, the vectors need to point to the new handlers a new area of store is chosen. The vector addresses are then stored there, and the base register contents are changed to point to the chosen area. The vector to use to enter the handling routine for this interrupt is found as described in the previous section. Simple interrupt schemes have one vector per interrupt line and full vector schemes have more. Full priority schemes use external hardware to supply the offset into the vector table on the data or address lines for the chosen interrupt lines.

So that is all there is to handling interrupts. The reverse operations are used to restore the interrupted state and so carry on with normal processing. The previous state is restored by unstacking or by a workspace switch-back to the previous workspace. Interrupts at this priority level are re-enabled but the action of this is held off. The program counter is

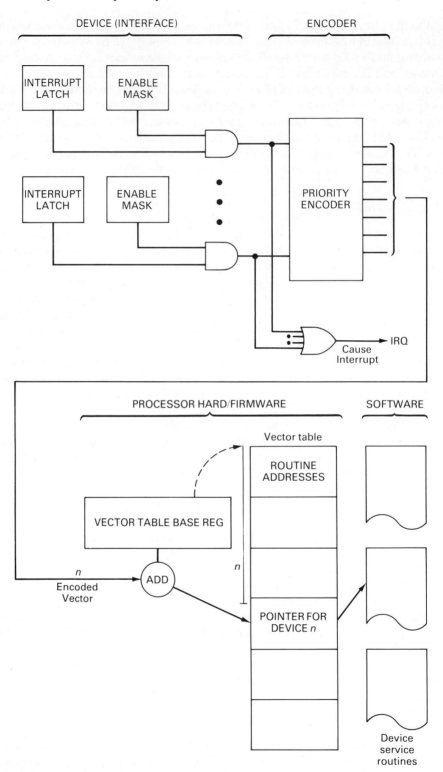

Fig. 11.8 Priority vectored interrupt

restored to the value it had when it was interrupted and the previous interrupt re-enable becomes effective. Commonly this is all done by an interrupt return (RTI) instruction. If an operating system reschedules tasks following an interrupt then the original program counter and state are not restored, but a switch is made to the new task's workspace and program.

There is one serious drawback though which creates problems for software writers and for users because of the unreliability and loss of speed it causes. It has been implicit throughout the discussion of interrupt handling that after an interrupt occurs there will be a handler routine to service it. This is not the case with common business and industrial micros. It is not that there is not a handler but there are **many** handlers for **one** interrupt. Consider, for example, a keyboard on a business micro. Different key depressions require different actions but all cause the same interrupt. Programs may be running for a word processor, some printing, a calculator and communication with a local area network, etc. Each program requires some user interaction and it all comes via the keyboard. This is easy—a handler is written which checks the key pressed and has a dispatch address table with an entry for each key to get to one of a small set of subsidiary handlers. But what if one of the processes is a mapping process? That is, on pressing a key e.g. [Function key 1] this handler changes it into another number e.g. [3] or worse into a sequence e.g. [H] [e] [l] [l] [o]. The handler now has to go back and pretend to be a sequence of interrupts (and characters) which may require different handling. The problem then is one for software writers who must adhere to some standard software interface to get out of the mess.

The situation is worse for the real-time clock interrupt. There is only one interrupt and it cannot be split up further yet it must service error timeouts, events to be activated at given times or after given delays, the time-of-day clock, updating displays, etc. So now we have a problem; the traditional solution has been to provide supervisor calls:

- **Return the address of the current handler**
- **Link this handler into the specified interrupt.**

The new handler then gets the address of the current handler (which is still wanted) and links to the interrupt. On interrupt, control is passed to the new handler which then does its thing and jumps to the old (current) handler. Eventually the final handler in the chain executes the 'return from interrupt' (RTI). This is slow and unreliable as a new and untried handler is put in **before** all the trusted ones. It is also plain messy.

A new instruction is really needed to return from the interrupt handler but also to check to see if there are any other handlers to be entered. This can be implemented using the 'software interrupt' mechanism which is an instruction which mimics an interrupt with its own vector etc. This is slow as the entire interrupt mechanism is rerun with all the saving and switching. If all handlers were written as stand-alone items each could be vectored to directly and return by a 'Conditional Interrupt Return'. This either does a real interrupt return or if a handler address remains to be serviced then jumps to it. It is faster to make this a 'vector link' which jumps to the next handler directly, and have a dummy handler as the last which just does the real RTI. This system has not appeared in commercial micros yet and they stagger on with chains of non-uniform handlers trying to coexist.

To recap on this chapter, the preferred methods of I/O handling are:

Store-mapped input/output which coexists well with Direct Store Access and Direct Device Transfer autonomous units when greater speed is needed. All I/O should be interrupt driven unless the transfer is atomic or autonomously atomic. In the first case the transfer has already occurred so no interrupt is needed. In the second case the interrupt only occurs at the end (or error) of the autonomous transfer. On interrupt state-saving should be by workspace switching or failing this by stacking. A priority encoded vector system should be used to determine which device caused the interrupt if there are many interrupting devices. A table in store holds the addresses of the service routine for each interrupting device. The resent vector and the other interrupt vectors should be located at opposite ends of the store of failing this a vector table base register should be used to point to the vectors addresses currently in use. Finally a coherent solution should be adopted if one interrupt is to be serviced by more than one handler. Ultimately all of this has to be mirrored up to the high-level language. This should provide for ordinary code procedures, and interrupt handling procedures that are linked to the selected device interrupt. It should use special functions (such as read or write) to specify I/O and the compiler should map these into store mapped or I/O mapped operations as applicable.

12

Integrated Interfaces

One side-effect of the development of very large-scale integrated circuit technology has been pressure to use only a few interface designs. A small number of 'standard', yet general purpose input-output devices have been produced. Many manufacturers have surprisingly similar approaches to the specifications for these devices. The flexibility is achieved in every case by making the chips programmable. A chip is connected to the processors' data bus, control bus and a vestigial address bus. Registers within a chip can be loaded to configure it to the exact requirements of the system. The options available are arranged to be as wide ranging as possible whilst keeping the chip simple. As well as more complex chips to cope with some international standard interfaces there are three generic choices. These are the parallel, serial, and counter/timer adaptors.

12.1 Programmable Parallel Interface

A general-purpose parallel interface chip, variously called a PPI, PIO or PIA, will try to provide as many input and/or output links as possible. It also includes some control lines for handling strobe outputs and interrupt inputs. The constraints of the chip package dominate this particular design. To interface to an eight-bit micro there will be eight data lines to connect to the bus. Read-write, timing, chip select, interrupt and address lines will leave only 20 to 24 lines for actual inputs and outputs from a 40-pin chip. Figure 12.1 shows the basic layout of the external interfaces and internal register structure of a typical PPI.

This particular one has two independent eight-bit I/O registers to hold output values. Inputs could be gated directly to the data bus for transfer to a register or store location but as the outputs also only appear transiently on the data bus then they must be held or **staticized** in a register (DATA). A second register is needed for each side to fix which bits (pins) are to be inputs and which are to be outputs (DIRECTION). Finally a control register is needed for each side. This sets the control lines as inputs or outputs, edge- or level-triggered, normal- or inverse-polarity operation and so forth. The control registers also enable the interrupt inputs through the chip.

Some interesting points arise from this design, particularly if the chip is to operate by store mapping. There must be a vestigial address bus connection (RS) to the chip in addition to the chip select (CS) line. These connections are to select which register we will access for a given read or write operation. The example has six registers hence 3 address lines would be needed. The direction registers, however, must be protected against any accidental write (e.g. one changing an output to an input) leaving an output undriven. A bit in the control register could be used as a 'write-protect' bit but there is a neater solution. We can **hide** registers behind other registers. If a control bit (e.g. DR) is set then one register is accessed; if it is cleared the other register is selected. This hidden mode of addressing is shown in Fig. 12.2. As the same address is used for both registers fewer address lines are needed; only two in this case. One initial reset access is set to the direction registers and the bits in it are set to give all inputs. Subsequently the directions can only be altered explicitly.

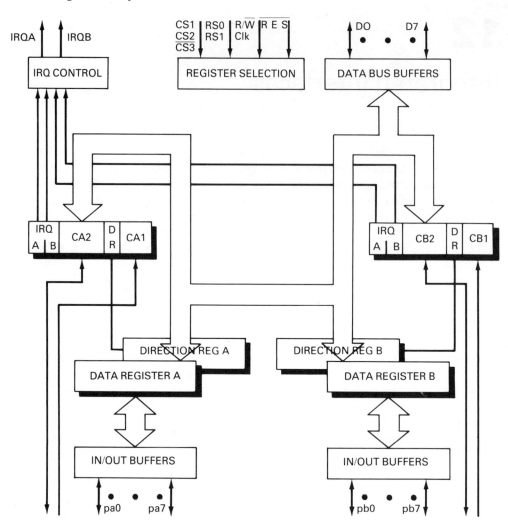

Fig. 12.1 Parallel interface adaptor

The control registers in this example contain the *hidden-access* bit, three bits for an interrupt input and four bits for an interrupt input or strobe output. For an interrupt input we need an enable/disable bit, a bit to indicate the state so that it can be checked and a bit to program whether the state changes on a rising or a falling edge. An extra bit is needed to select the I/O control direction. If set as an input it is broadly as already described. When chosen as an output it may either be used as a straight output bit (mode 10 sets output = 0, 11 sets output = 1) or else as a strobe. The two sides A and B are arranged differently in this case. The A-half has the strobe go low on the clock signal following a **read** on the associated data register and stay low until the next clock (mode 01) or the next interrupt input (mode 00). The B-half has the same operation but the strobe follows a **write** on the associated data register.

A number of simple operational optimizations can be made by observing the likely sequences of interactions between a simple microprocessor and such a chip. Once an

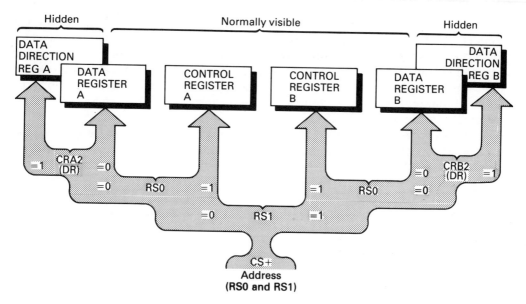

Fig. 12.2 PIA addressing and register protection

interrupt has been caused, in one of the ways described, it would be likely that a read would be performed on the associated control register to determine the cause. Reads might be necessary on more than one control register for this purpose and so it would not be safe to have this action clear the interrupt bit(s). It is also reasonable to suppose that if an input operation were being requested by the interrupt, that a read operation would subsequently be performed on the associated data register. It is perfectly safe to have this clear the interrupt and it saves having explicit instructions to do it. A similar arrangement could be arranged for a write side. A simple observation of the arrangement of the register select address lines shows that they can be wired to give an ordering of Data-A, Data-B, Control-A, and Control-B. This would allow sixteen-bit transfers to be carried out with one instruction (via an index register). If the more usual arrangement of address lines was retained but the least significant bit was inverted before connection to the chip then a sixteen-bit transfer of Control-A followed by Data-A would be possible. This would read the control register, read the data and clear the interrupt flag(s) all in one go!

It is apparent that a very simple part provides great flexibility because it is programmable. Other general-purpose interface chips have different modes but all are combinations or variations of the options discussed.

12.2 Programmable Serial Interface

A general-purpose chip can be made to do more for us if more is known about the end interface. A range of chips is made to interface to the RS232C (V24) standard described in Chapter 8, and called variously Universal Asynchronous Receiver Transmitter (UART), Asynchronous Communications Interface Adaptor (ACIA) or Serial Input Output (SIO) chips. The interface to the microprocessor's bus is the same as for the parallel interface: data, control and the vestigial address bus. The interface to the device will be serialized data for transmission and reception along with drivers and receivers for some of the control signals. From the standard it can be seen that three required control signals must be present. Request to Send originates from data terminal equipment (DTE) while Clear to Send and Data Set Ready are inputs to it. Some chips make this a symmetric set by adding an originating Data Terminal Ready. The RS232C interface can be operated synchronously (e.g. with clocks on pins 13 and 15) or isochronously. The chip will have to be set up for the required transmission mode and will also need suitable clocks. These could be generated externally and supplied to the chip via two input pins, or counters could be included to divide down the micro-supplied clock to the chosen values. Either method must be completely variable as there are no defined speeds of operation in RS232, only some dozen or more commonly used values. A bit will be necessary to specify which type of synchronization is to be used. If the baud rate is generated on chip then eight or more bits will be needed to set up the master clock dividers for each of transmit and receive speed. A special case (e.g. 0000) could be used to select synchronous, externally clocked, transmission. So there would be selectors for:

Isochronous ($\times 16$ clock) or **Synchronous** ($\times 1$ clock)
and/or
<Transmit rate selector><Receive rate selector>

There are no standard definitions for word length, parity and framing. The word length could be five, six, seven, eight or even nine bits. The shorter lengths have fallen from favour, the old five-bit Baudot code only being used on outdated telex machines. Consequently data usually occupies seven or eight bits. Parity is often added and checked on reception automatically. This is a simple job requiring only a few exclusive-or gates. Parity can be odd or even, a bit being added to leave an odd or even number of zero bits. A parity bit can be **marked**, always set to be a one, or **spaced**, always set to be a zero. Both of these are a complete waste of a bit for each character transmitted. Their sole purpose is to pad out a seven-bit ASCII coded character to fit an eight-bit data frame. Finally one may not bother with parity at all, leaving the data bits to be transferred transparently. The isochronous transfer mechanism has a start bit preceding the frame and a stop bit to follow it. It may have more, say 1.5 or two stop bits. There was a historical reason for the two stop bits case. Original teletypes had their timing determined by a motor and clutch arrangement. Whilst running at ten characters per second they could start up in one bit time but took two bit times to come to rest. Hence one of the 'standard' speeds is still 110 bits per second of eleven bit frames. Thus a typical set of eight **modes** might take three bits to specify and include:

7 bits + even parity + 1 stop (9)	8 bits + no parity + 1 stop (10)
7 bits + odd parity + 1 stop (9)	8 bits + no parity + 2 stop (11)
7 bits + even parity + 2 stop (10)	8 bits + even parity + 1 stop (11)
7 bits + odd parity + 2 stop (10)	8 bits + odd parity + 1 stop (11)

As the serialization and transmission of data bits are often very slow compared with machine speed it will be essential for our serial chip to provide interrupts when it is ready to transmit, on completion of reception of a character or if an error is detected. It is obvious that an interrupt should be raised when the transmit buffer is empty and another character may be transferred into it to be serialized and output. It is also apparent that an interrupt should be raised when the receive buffer is full after a character has been received. This should not occur until after the character has been deserialized and the data must be taken. When a character is received it will be checked, by the chip hardware, for errors. These will be parity error according to the mode settings, framing error according to the stop bit settings and overrun error if a character is received and is to be transferred into the buffer but the previous character held there has not yet been taken. As these errors will always be associated with receipt of a character a separate interrupt is not required, but it will sometimes be desirable to handle errors in a different module. In these cases a separate interrupt would be used but most serial interface chips do not have this. The final case for interrupt support is the set of interface control lines. If the DCD or CTS lines change, how should the system be informed. This brings in a larger question of the linkage between **control** lines and **data** transfers.

On the transmit side we need to enable or disable the interrupt, raise or lower RTS and handle the setting of the rather obscure, but very useful, **Break** condition. The Break condition is a start bit (spacing) which persists for an arbitrary time and so it must be explicitly started and stopped. It does not start and stop like a transmitted character so the concept of an interrupt on its completion is meaningless. Also we will not want to service an interrupt after every character-time of the break, so implicitly the interrupt must be disabled. We can have all possible patterns of these or provide an encoded form of just those which are legal or useful. For example:

RTS false + interrupt disabled   RTS true + interrupt disabled
(no transmission requested!)   RTS true + interrupt enabled
                                       RTS true + break condition

On the receiver side we will simply need to enable or disable the interrupt. The cause will have to be determined by subsequent inspections of a status register containing the Character In the Receiver Register, DCD and CTS bits. Some chips link the incoming control signals (from a modem) to the transmit and receive mechanisms. We have the choice of having them as just input bits we can test, or as bits which cause interrupts on changes of their condition. The alternative mechanism is to use Clear To Send to stop transmission. When character frames are sent isochronously a character frame is itself the minimum unit we can send so we must not stop until the end of the current frame. Some chips are wrongly specified in this respect and can output partial character frames. Similarly DCD can be used to stop reception of data, or to stop setting the flag or interrupt after a character is received. Such attempts at hardware **flow control** have not met with much success as they are not standardized. Software attempts using particular patterns to stop and start data are more flexible but are still not standardized. There are at least four types in common use.

In the simple form of serial interface chip there are four registers. Again this would imply two register-select lines but some smart observation reduces this to one. There are two data registers: one to hold the data to serialize and transmit and the other to hold the received data after deserialization. The first is write-only, the second read-only. There are command and status registers having, typically, the contents shown in Fig. 12.3. The first can be write-only and the second can be read-only. Thus a single register select and the read-write line can allow access to the four registers. The more complex versions of serial interfaces include more register selects to get access to baud rate dividers, extended control signals, etc. The complete structure and connection of a typical chip is shown in Fig. 12.4. There are many varieties of this type of chip as each manufacturer takes different choices. The basic interface for RS232 to operate in isochronous or synchronous modes with off-chip clocks is seven lines. These are: transmitted and received data, transmit and receive clocks and the three control lines RQS, CTS and DCD. So a single interface fits in a standard 24-pin chip. A dual interface with extended control lines, on-chip baud rate generators and more command options easily fits onto a standard 40-pin chip. In fact, if only the much more

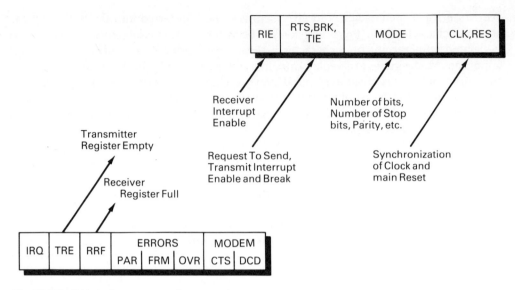

Fig. 12.3 Serial interface status and command

common isochronous regime is used with on-chip clock generation, four minimum required interfaces fit in the 40-pin chip. Of course higher levels of protocol could also be included and such chips are described in Section 12.4.

A final point of note on the simple serial chips is that even with chips as easy to specify and as logically simple as these there will still be errors in specification or maladroit design. As an example, in a local area network it was decided to use one of the most popular serial chips and phase encode the output. The encoded data was transmitted over the network and the data and clock were recovered by a simple decoder. Thus each data bit had a clock in mid-bit to clock it into the receiver synchronously. Having successfully clocked in the start bit, data bits, parity and stop bit, with the same modes set at either end, one would have reasonably expected that the character would be transferred into the receiver register. No such luck! The processor has to wait till the **next** clock edge to get the data. There is nothing in the documentation even to hint that in the one-clock-per-data-bit-synchronous mode you need one extra clock to get a character in. The moral is obvious: don't trust data sheets to tell the whole truth. A further example is an ACIA marketed with the unpleasant quirk of throwing away characters received if the requested parity check is incorrect. For a screen display it is visually better to see a correctly aligned page with any errors highlighted rather than throw a section of the screen out of alignment. In about 10% of cases it will be the check that is in error, not the data bits, so it is far better to pass on the characters with status bits to indicate any supposed error.

12.3 Programmable Counter/Timer Interface

The third of the triumvirate of common integrated circuit interfaces is the counter/timer chip (CTC). There are a multitude of counting, timing and pulse generation jobs for which a microprocessor is excellent but is limited in speed by its instruction cycle time. Adding very simple external hardware in the form of a CTC extends the rate well up into the megahertz range. The addition of a prescaler before that gets the rate up to a hundred megahertz and

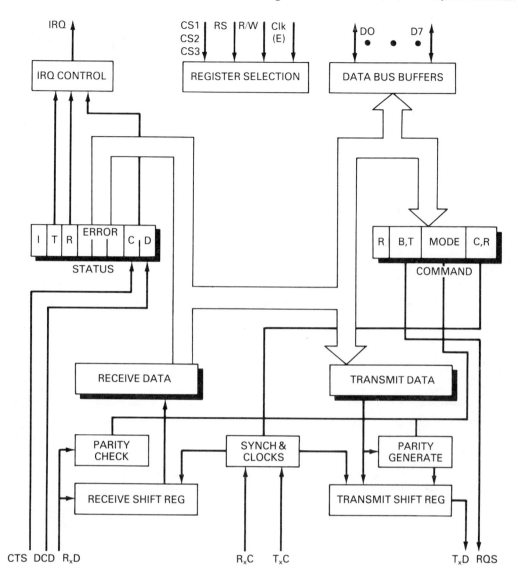

Fig. 12.4 Serial interface adaptor for RS232C

above. A general-purpose counter/timer will provide both inputs to count and measure, and outputs to generate desired frequencies or pulse widths. The content of a typical chip is shown in Fig. 12.5. The micro interface, not surprisingly, is very similar to those of the chips already described. Because the CTC has more internal registers there are more address pin inputs for selecting between them. Typically a CTC has three or more 16-bit counter registers with their associated (but hidden) latch registers. Extra buffer registers have to be included to permit 16-bit values to be loaded into the chip by two separate 8-bit transfers, but then be transferred to the latches at the same instant. Similarly when the contents of a counter are to be read, two transfers are needed. Again one 'half' must be held in a temporary buffer, while the other half is read, to ensure consistency. The choice of which to

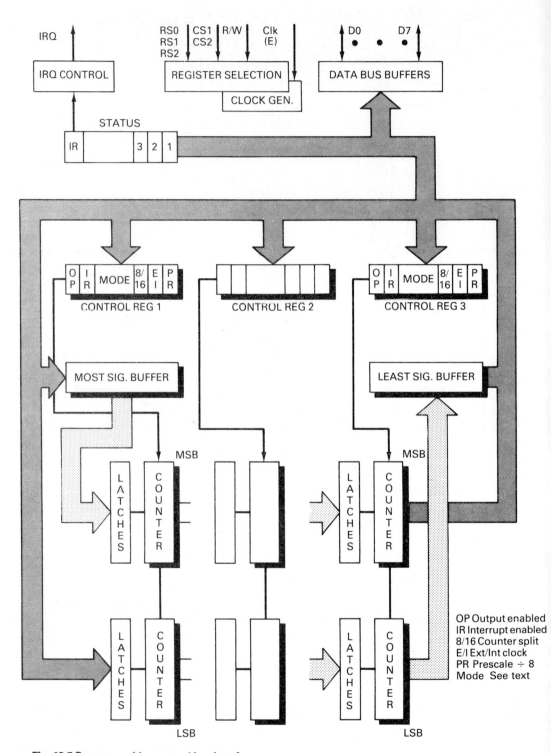

Fig. 12.5 Programmable counter/time interface

read first is fairly arbitrary but as the lower half is more likely to change then it can be held while the top half is read. The opposite can be done when a write into the chip is performed. The top half of the 16-bit word can be loaded into a buffer for transfer to the latches at the same time as the lower half is loaded directly. Of course, if the chip had a 16-bit parallel interface this problem would disappear but eight more pins would be needed! An interrupt line, system clock, read-write, reset and power lines are all required and we would be left with only six pins from a 24-pin chip. Thus 28-pin or 40-pin packaging is used. Each timer/counter section included needs three external pins, an output plus clock and gate inputs.

Each counter/timer also requires a full 16-bit counter, 16-bit latch register, the buffers discussed above, some status bits and a command register to choose one of the many operating modes. Obviously **input** or **output** operation must be selected. The same basic counter can be used to generate output patterns or to measure inputs presented to it. For output either **continuous** or **single pulse** signals could be produced on the output pin. These can be generated relative to the **internal** (system) clock or relative to the signal on the **external** clock pin for the selected counter register. Continuous (repetitive) outputs can be **square waves** which need a single counter to set their frequency or have a **variable mark to space** ratio. This needs two counters and so either two of the 16-bit counters are linked and used together or a special mode is implemented which splits one 16-bit counter into two 8-bit counters to set the mark and space periods. The range of available frequencies can be increased by letting one of the 8-bit counters set the mark period but use both (i.e. 16 bits) to set the frequency (mark plus space). This is more versatile and easier to make as the carry after the eighth bit of the 16-bit counter is simply tapped-off. The output can be controlled internally, or by the external **gate** control, or by both.

Single pulse operation, as its name implies, means that only a single cycle of one of the continuous patterns is actually output no matter how the internal counters are then stepped. A special case (count = 0) can be used to set the output to a fixed low or high state when pulsed output is not needed. One further detail of output modes is that we may wish to arrange an interrupt to be caused at the end of each counter cycle. This interrupt function will need to be enabled or disabled and it will also be useful for all the circuitry to be used to cause the interrupt, but not actually to output the pattern that has been generated. Thus command bits are needed to **enable** the **interrupt** and the **output driver** separately.

There are a couple of choices for configuration and some operating modes to select from to set up the measurement of inputs. If the internal (system) clock is selected then we will measure transitions of the gate input relative to it. If we use the external clock input then we can count it directly or with respect to changes on the gate input. This simple configuration mechanism gives enormous flexibility in operation. Command register bits select between comparisons of **frequency**, i.e. complete cycles, or **pulse width**, i.e. marking or spacing time. In both cases comparisons are made between events on the gating input and periods set by the internal counter. The comparison can be chosen to be **greater** than or **less** than. So, for frequency measurement for example, an interrupt may be generated if the gate input changes from one to zero, the counter runs its allotted time and the gate input has not changed from one to zero again in that time. The interrupt could be set to occur if the input did change from one to zero within the set time. Obviously for pulse width measurement opposite, rather than similar, transitions are acted on. By linking separate counter sections together very rapid measurements can be made, using one counter to set a system clock dependent output during which time an unknown input frequency or width is counted. If the frequency range of desired inputs exceeds that which the chip's counters can handle an on-chip prescaler can be selected. This is a special input capable of handling higher frequencies and dividing them by a fixed number (8 or 16 say) before passing the transitions into the ordinary counter. If even this is not fast enough then an external prescaler (divider) must be added into the signal path before connecting it to the counter/timer chip. It is very likely that a different technology would be used in the prescaler chip to let it operate at the (much) higher frequencies.

12.4 Complex Integrated Interfaces

The three types of integrated interface discussed so far are relatively simple. They are easy to design using only large-scale integration and are easy to understand and use. They can also be fairly rigorously tested so with only minor exceptions should be free from quirks.

It has been noted in earlier chapters that it is highly desirable to make interface connection simpler by moving as much of the function as possible into logical rather than physical levels of the interface. With the advent of very large-scale integration it has become possible to include all of the logic for the most complex interfaces onto a single chip. Numbers of such complex integrated interfaces have been produced for the obvious standard interfaces. The only cautionary note for their use is that interface standards are not as 'regular' in their design as, for example, microprocessors. Therefore they are more prone to design faults. Their manufacturers are also somewhat reluctant to release 'bug sheets' which detail all the problems or restrictions found in a chip when used. The prevalence of 'knocking copy' advertising in some countries accounts for some of this, but whatever the reason it is a nuisance for us, the users. However, with this proviso, these chips are a great saving both in design time and production cost. Chips are available for the functions listed below, among others:

- **IEEE 488 Interface**
- **Binary Synchronous Control serial interface (BSC)**
- **High level Data Link Control serial interface (HDLC)**
- **Cathode Ray Tube Controller (CRTC for visual displays)**
- **Raster Graphics Controller**
- **Data encryption/decryption chip (to NBS)**
- **Analog interfaces (A–D, D–A)**

Chips are also produced for a number of complex system support functions. These simplify design and reduce chip count and system cost. Examples are:

- **Direct Store Access control (DSA)**
- **Interrupt prioritizer and vector handler**
- **First In First Out buffer for data rate smoothing (FIFO)**
- **Dynamic store refresh controller**

The main restrictions on all these interfaces are pin-count and speed of operation. As the functional operation of most of these is described elsewhere in this book, here we can concentrate on the particular requirements for integration.

The IEEE 488 interface chip needs the usual data, control and vestigial address connections to the microprocessor. The IEEE bus has 16 lines, 8 bidirectional data and 8 control. For the reasons discussed in Chapter 10 the high current drivers are located on separate chips and so extra lines are needed to enable them for the correct direction (reading or writing). At least two signals are used for this. Also as the interface may operate at high speed it would be sensible to include the small amount of extra circuitry to handle direct store access. This only needs two lines, **request** and **grant**, if a full DSA address mechanism is included for other devices as well. Thus 40-pin packaging will suffice. Most of the available chips support all three operating modes: i.e. talker, listener or controller, and need a number of internal registers to set up their addresses, operating mode, parallel poll response etc., in addition to those for data transfer.

Serial synchronous (binary synchronous) communication chips are quite similar to the ordinary 'asynchronous' ones but with various additions to handle the first logical level of communication. They handle byte-organized protocols and can usually be relied on to generate and strip synchronous idle (SYN) characters, to retain synchronization, and to insert and remove data link escape (DLE) characters to allow data transparency. The patterns for these characters are sometimes loaded into registers rather than being fixed in the chip. The generation and checking of block-check characters (longitudinal parity) may also be included to reduce marginally the computational load on the processor. Compared to the simple 'asynchronous' chips there will be more modem control lines, hence 40-pin packages, and a deeper level of buffering for transmitted and received data to smooth the interrupt handling load on the processor. BSC communication has, however, been overtaken by the more powerful bit-oriented protocols and the chips to support them.

High level Data Link Control (HDLC) chips support the modern bit-organized, position-dependent framing and transmission schemes. They can synchronize to the data and then detect the flag pattern (01111110). They can then extract and control patterns before buffering data. The chip both inserts zeros into transmitted data and strips them from received data (after five ones in a row) to ensure transparency in all fields other than flags. There is obviously yet more data buffering as these chips are designed to operate at rates in excess of a million bits per second. There is, however, a reduced requirement for modem control lines as HDLC is generally operated over the X21 style of physical interface. The whole data frame including address, control and information is checked using a Cyclic Redundancy Check (CRC) which produces 16 bits of redundancy. These are automatically generated and added to the end of the information which is transmitted prior to a flag sequence being sent. On reception after bit stripping the CRC is again generated automatically and is checked with the pattern received. This CCITT V41 standard CRC detects over 99% of all possible burst errors. Chips have already emerged to handle the next logical level above this, but the higher levels of communication are still subject to some modification of their standards.

Two further useful extras on the serial communications side are FIFO buffers for changing the rate at which service needs to be provided by the processor, and data encryption/decryption chips. The former are used when the level of buffering provided by a communication chip is low enough to cause an excess processing load or, worse, lost data. They are also used to provide fast and flexible links between processors in multi-processor systems. The latter are used to provide secure communication. This is secure in the sense of private rather than secure in the sense of reliable. They are almost all based on the US National Bureau of Standards encryption system (DES) and some are not available outside the USA. It is likely that better and faster chips will become available soon to serve this need, particularly in the banking and finance areas, for secure communication.

Many of the complex interfaces can be implemented by an interface to the microprocessor bus, an interface to a device and some processing function in between. This description fits a single chip computer just as well. Consequently general-purpose complex interfaces can be made with only a small customization of one of the better single chip microcomputers. This has not escaped the notice of manufacturers. It is a cheaper and more reliable route than a custom-designed chip and so a range of 'integrated interfaces' can be found which are no more than single chip microcomputers with suitable programs fixed into read-only stores inside them.

13

The Choice of Microprocessor

And so, somewhat tardily, we come to decide on which machine we should base all our work. This choice will be limited in many companies to a particular range of hardware because of previous projects, available support equipment and expertise, or simply executive directive. If there is no restriction you simply buy the best you can afford. If it is a large system then the processor is a small part of the cost, so buy the best. If the system is virtually a single chip system you still buy the best but from an entirely different set of processors. The range of processor power available is vast, running from 4-bit machines, little more complex than logic controllers through to 32-bit processors more powerful than many minicomputers.

There are two approaches for manufacturers producing new processors: **evolutionary** or **revolutionary** design. The problem with evolving designs, which retain compatibility with previous models, is that early structural mistakes are turned into the established yet unsound features of the new range. The process of evolution of microprocessor families has meant that systems which **were** bad because of technological restrictions are developed into systems which **are** bad, despite technological advances. The market is split between companies which vie to be always first with a new product and those which are somewhat later with theirs but almost invariably produce cleaner, better designs which are easier to use. Retaining upward compatibility in new processors may be likened to having the same tyre sizes and chassis members from the 1940s Ford Popular car in the current model bearing the same name!

So, how do we choose a processor to use? We can start with the users' viewpoint and considerations:

- **Cost of micro**
- **Reliability**
- **Performance** (processing power and clock speed)
- **Flexibility** (ease of use)
- **Complementary hardware** (ease of interfacing)
- **Environmental constraints** (power consumption etc.)
- **Manufacturers' support** (second-source suppliers)

The cost and manufacturers' support arrangements should be easily found. The availability of support chips, such as those described in Chapter 12, should also be easy to ascertain. The other factors require some study of the manufacturers' data sheets. The power consumption and other environmental factors will have been measured and be quoted with ranges for typical devices. The reliability will be estimated from a set of accelerated tests at higher temperatures and voltages. A confusion arises with the clock speed as some manufacturers quote the frequency of the crystal oscillator while others quote the bus access clock rate. Thus a 4 MHz Zilog Z80 is significantly slower than a 2 MHz Rockwell 6502. To evaluate performance and flexibility fully we need to look beyond, as well as at, the data sheet.

13.1 Microprocessor Architecture

Processors are usually categorized by their data word length (width!) but this is getting to be rather misleading. Should it be the bus width, the register width or the width of the arithmetic and logic unit we count? It certainly cannot be the instruction word length as most microprocessors now have a variable instruction length. There is a range of characteristics we may use as a measure of the processor architecture's quality, power and ease of programming:

- **Word length**
- **Register structure**
- **Bus structure**
- **Instruction set**
- **Addressing structure**
- **Addressing range**
- **Addressing modes**
- **Data types**

There are, however, a number of conflicting constraints which limit a designer's ability to deliver the optimum mix of these architectural features. The implementation technology determines the number of logic elements which may be integrated onto the chip. This number is determined by both the minimum feature size (in microns) and the power dissipation of the chosen technology. The bus structure and to some extent the word length are limited by the number of pins the package can contain. Finally the performance is ultimately determined by the speed of the individual logic elements. This is fixed by the chosen technology and by the size of each element. Up to a point, the smaller they are the faster they get, as do the interconnections between them.

Having said that the word length is not a good guide to a processor's quality or performance, the next thing the manufacturer will show on the data sheet is the register structure. The reason is simple—you can make a nice drawing of it! If there is an adequate set of registers and they are treated uniformly then there is no problem and the drawing adds no information. If one of each required type of register is included and each is treated correctly then the same situation prevails. If there is an arbitrary definition of special-purpose registers then there is a problem but the drawing is of no help in overcoming it.

Commonly a drawing of the chip and its pins will be shown next. The bus structure is only of real interest for the data bus width, the address bus width (hence its immediate range) and if the two are multiplexed to share pins or not. Multiplexing saves pins on the package but is slower and provides no simplification of design because store chips and I/O chips do not have multiplexed lines. A more sensible approach to cutting the pin count would be to adopt directly the address high/low multiplexing arrangement of store chips to simplify their connection.

So we come to the characteristics which really matter and we may as well start with the instruction set. It can be **normal**, **enlarged** (or complex), or **reduced** in size. It should also be **orthogonal**. Orthogonality applies equally to the whole architecture. It means that anything which has no logical connection should not be connected or mixed. For the instruction set this means that the function or operation code is separate from the operand address, from how the operand is accessed (the addressing mode), and from what the operand is (its data type). Sadly many micros are not orthogonal and make life unreasonably difficult for compiler writers, and inefficient for users.

Instructions (*Op-codes*) are needed for arithmetic operations, logical and shift operations, data control and movement, program sequence control and some housekeeping functions (interrupts etc.). In a 'normal' set there are perhaps fifty different basic instructions. Some of these are variations on a theme; for example, there may be arithmetic and logical shifts and rotates in both directions even though not all are necessary. The enlarged or complex instruction sets increase this number to some hundreds. The snag is that although there is always just the right instruction available, it is impossible for a compiler (or compiler writer) to be clever enough to match all the special cases. The cost we pay is in performance. The greater complexity of the instruction decoder slows the operation of the whole machine. At the opposite end of the spectrum are the reduced instruction set computers (RISCs). The instruction set is simplified to less than 20 instructions by only including necessities. The architecture is then much faster as all the simple instructions are quicker. Complex operations are built from sequences of simple instructions. It has been demonstrated that this approach gives both more compact code (less store) and faster code in the majority of applications. It is also much easier for a compiler to optimize the code it produces. The examples later in this chapter all have 'normal' sets with the exception of the transputer which is a RISC.

Our set of instructions with their op-codes filled in now have to access their operands and this brings us onto the addressing structure. There are three aspects to this: the **structure**, the **range** and the **modes**. The addressing range, the number of different locations we can address, is determined by the address word length and the size of elements we wish to address. Figure 13.1 shows an operand address and how its length and range vary. Early micros only had sixteen bits of address and only addressed bytes (eight-bit values). Thus they had a range of 64 kilobytes (65 536 bytes). Obviously they could address only half this number of sixteen-bit words and fewer still instructions which averaged more than two bytes each. This range is still adequate for simple microcontrollers but for personal computers, workstations and complex control functions is too small.

The next obvious address length was 24 bits (three bytes). Many of the good 16- and 16/32-bit processors used this as a reasonable compromise. It gives a range of 16 Mega (16 777 216) elements. This has proved adequate for real store sizes but not for virtual store, including backing store. It is definitely inadequate if we have a uniform, object-oriented architecture where 'files' or backing store are treated the same as data structures in real or

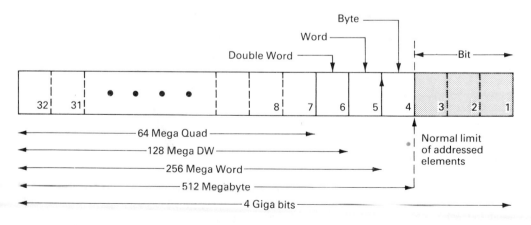

Fig. 13.1 Address length and range (fully decoded)

virtual store. Then a fully uniform space is needed and 32 bits, the obvious choice, gives 4 Giga (4 294 967 296) elements. This should be adequate to hold everything one could create or use at present, but who knows for the future, perhaps 64 bits?

The problem with adopting such long addresses is that they must be stored even when we only need a short local address. This is overcome by having the full-length address available for every instruction, but permitting a reduced length whenever the compiler decides it is all that is needed. A single bit extra is all we need to choose 16 or 32 bits. Two bits would select 8-, 16-, 24- or 32-bit addresses or offsets.

If we have such a large address space there is a great temptation for a microprocessor designer to put some **structure** on it in the architecture (Table 13.1). Any artificial structure, such as splitting it up into fixed size segments (say 64k each) is bound to be a mistake. It arbitrarily limits or constrains the user to a particular approach. Some manufacturers have done this only to retain compatibility with a previous processor. Consider, for example, a simulation program which needs a large, randomly accessed data structure. If fixed segments are in operation almost every access will be to a new segment. This needs action from the system to change the segment base register to the new value and has a drastic effect on performance.

Any address structure must be decided on by the system designer, not the processor designer. A linear address space can then be paged to allow a reduced real address space to run a larger virtual one. Variable length segments may then be used to improve inter-task protection.

Addressing modes, on the other hand, are an integral part of an architecture and greatly influence the power of a processor. There is a wide range of modes and they are not always given the same names. They can be simply categorized though as in Table 13.2.

The first part of the mode determines how immediately the operand can be found. **Implied** addressing, sometimes called inherent addressing, means that it is implicit from the

Table 13.1 Addressing structures

Desirable	Restrictive
Linear	Linear bank switched
Linear-paged	Blocked
Linear-paged segmented (variable length)	Segmented (fixed length)

Table 13.2 Addressing modes

A	B	C
Implied	Absolute	Index register
Immediate	Relative	Base register
Register		Program Counter
Direct		Index and Base
		Store location
Indirect ⎫	A/C combinations	⎧ Pre/Post index
Multi-indirect ⎭		⎨ Auto increment
		⎩ Auto decrement

operation code where the operand is located. Some stack operations, such as drop or duplicate, and operations on the status word are of this type. **Immediate** addressing locates the operand in the rest of the instruction word in place of an address. This mode is very convenient for loading constants. **Direct** addressing, which is perhaps the most obvious form, contains the address of the operand in the instruction. Thus an extra store access is required to reach a variable in read-write store. **Indirect** addressing contains an address in the instruction word. This address when accessed supplies a value which is itself an address: the address of the operand. Thus we can modify the second address to, for example, step down the entries in a list or table. A slight variation on this was to locate the second address in a register, but used in the same way. This has only a slight speed advantage at the cost of fewer possible indirect addresses available at a time.

These modes are all that were found on early minicomputers, however they are not adequate to support high-level languages efficiently or to allow for position independent code. A small extension was to allow more than one level of indirection.

The real improvement comes when instead of using **absolute** addresses as in all the previous modes we use **relative** addresses. This means that a computation, usually addition, is performed between the otherwise absolute value we had and some other value. The simplest case of this is to use a register as an **index** register. With direct indexed addressing the address in the instruction is added to the contents of the specified index register to give the address of the operand in store. Manipulation of the index register is faster than manipulating store. We can also arrange for automatic incrementing or decrementing before or after an indexed access to step on to the next element of a structure. Using an indirect address combined with indexing we get two choices: do we index before we indirect (pre-indexing) or do we perform the indirect access to find the address from store and then add the index register to it (post-indexing)? The former is more convenient for accessing a dispatch table, for a CASE statement for example, while the latter is good for accessing arrays of arrays, etc. There is one point to note for many eight-bit micros, and that is the length of the address versus the length of the index register(s). Some micros do not provide both to the full length of the address. It is more important to have the index register of full length. Another restriction we will see more of later is that not all operations may be available on index registers. Sometimes only increment, decrement and compare are available and this is very inconvenient if elements in an array or record are not of unit or uniform length.

The second useful register for relative addressing is the **program counter** (PC). If an address from the instruction is added to the PC then operands are located relative to the program. This is obviously useful for transfer of control (branch) instructions. The need for program-relative branches is fully discussed in Section 14.2.

Making all accesses relative to some **base** register allows us to check and protect store accesses. It also simplifies access to global and local data structures. The combination of a base and a subsequent index is versatile indeed. Finally we can note that all of these address modification calculations could use store locations rather than registers with the exception of the program counter of course. This is more general though it requires a longer instruction to contain the addresses and is slower. Thus it is normal to use registers for all base, index and relative address modification.

Until recently the most undervalued part of the architecture of microprocessors was the range of **data types** and the manner of their support. Many analyses of programs have come to similar results about the most used data types. They come in the frequency order shown in Table 13.3.

Table 13.3 Data types listed in frequency order

1	**Address** (pointer)	16–32 bits
2	**Byte** (character)	8 bits
3	**Count variables**	<256 (many)
	Count variables	256 < x < 65 536 (few)
4	**Arithmetic variables**	4–36 bits integer or real
5	**Binary types**	8–10–14 bits
6	**Booleans** (semaphores)	1 bit

A reason for listing these is to point out why 8-bit micros have not only been popular but will remain so for many tasks. A more important reason is that **Address** tops the list. The most popular simple microprocessors, the Zilog Z80 and Intel 8085, and 8086/8 do not fully support the address type, though more modern designs do.

To ensure orthogonality in the architecture each instruction or function should be able to operate on every type of data. There is no reason why bits, bytes, binary coded decimal numbers, integers, double length integers, floating point numbers, etc., should not be added to another of the same type. A compiler can easily choose the **add** function, the correct data type and the addressing mode if all are separate. Thus all instrucions of a microprocessor should have the form shown in Fig. 13.2.

Of course one should also look at the I/O structure when considering an architecture but as discussed in Chapter 11 it can be added on by suitable chips, unless a 'single chip' design is required.

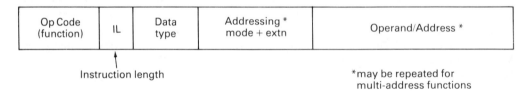

Instruction length

*may be repeated for multi-address functions

Fig. 13.2 Microprocessor instruction format

13.2 Example 4-bit Computer: TMS1000

Four examples of processor architectures follow to show the benefits and limitations of different chip complexities and word lengths. Though one-bit processors exist, normally used as very simple controllers, the smallest processors in widespread use are four-bit organized.

A typical and very common example is the TMS 1000 manufacturered by Texas Instruments Inc., and with versions selling cheaply in quantity. The TMS 1000 is one of a range of processors to a common architecture but differing in implementation technology, size of internal stores, number of input/output connections etc. As it is such a simple device many parts of a perfect architecture are omitted or reduced which paradoxically makes the device harder to program. As it is a complete single chip microcomputer it has input and output lines rather than control, data and address buses. The read-only store for program storage is typically a thousand or two eight-bit words and with the 64 four-bit words of read-write store for variables are internal to the chip. The read-only store for the program is masked at production time to fix its values so an emulator is necessary to test out programs before they are masked in. This is described in Chapter 14. One of the many package variants has 28 pins and its architecture is shown in Fig. 13.3. It needs two power supply pins and one for power-on initialization (reset). Two pins are required for the oscillator components to determine the clock speed and this leaves 23 pins for input/output.

These are arranged as four inputs which may be tested or read in, eleven latched outputs (R) which may be individually set or cleared and eight outputs connected from a mask programmable logic array. This arranges the decoding from five output bits which are loaded in parallel to the eight outputs. They can be easily used for driving calculator-like displays such as are also found on petrol pumps etc. The PLA connections are also masked at production time so testing is a slight problem. There are no interrupts, all testing being polled. The cheapest version of this chip clocks at 400 kHz giving instruction times of typically fifteen microseconds. This is quite adequate for dedicated control in consumer goods such as washing machines, toys, etc. There are only four inputs, but of course only four can be read

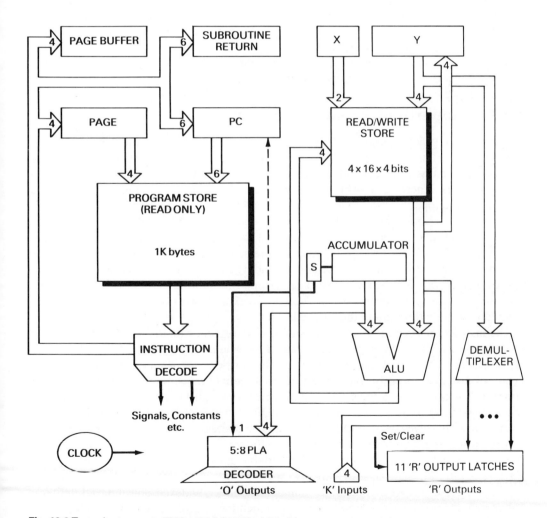

Fig. 13.3 Texas Instruments TMS 1000 (simplified circuit)

and stored at one time as it is a four-bit machine. The R outputs are used to multiplex (select) many sets of inputs to these, four at a time.

The RAM address needs six bits so two registers X and Y are needed to compute this. Full control is available over the four-bit Y register but only limited control may be exercised over the top two bits of the X register. Similarly the ROM holding the program needs a ten-bit address and this is only supported by split registers. On powerup the program counter is reset so that program operation begins at a fixed location. The ROM is split into 16 'pages' of 64 bits each and again full control is only available within a 'page'. The top four bits of address may only be changed by loading a constant to the page buffer register. If a call or branch is taken then the program counter is directly altered and the page address register is loaded from this page buffer. Only a single nested subroutine call is possible as the return address is held split between the six-bit subroutine return and four-bit page buffer registers.

Despite the many restrictions, and because of its low cost, the TMS 1000 has probably been the best selling computer of all time. They are found in a very wide range of applications from toys to industrial controllers and microwave ovens to petrol pumps. There is obviously an equally wide range of applications for which such simple architectures are not suitable. The next size up is able to provide a more complete architecture though a surprising number of examples fall short. The chosen example is one of the best of the eight-bit micros and was designed to be an eight-bit 'clone' of the Digital Equipment Corporation's 16-bit PDP-11 minicomputer.

13.3 Example 8-bit Processor: M6809

The Motorola 6809 has become very popular, particularly with those who appreciate good architectures. Figure 13.4 shows that it has a basic eight-bit structure, arithmetic and logic unit and data paths. The store address is 16 bits long as are the two stack pointers and two index registers. There is no arbitrary restriction on addressing or the mix of read-write, read-only or mapped I/O storage. The register structure is a good example of the 'have one of everything' school. There is a 16-bit program counter and an 8-bit status register. It has two 8-bit accumulators, but these may be treated as a single 16-bit register by concatenating them together. There are two 16-bit index registers. This is because to move data from one structure to another we need source and sink addresses to increment. There are also two stack pointers, one for the system stack (interrupts and subroutines, etc.) and one for the user stack (data, parameters, etc.). The two index and two stack pointer registers are treated uniformly throughout the instruction set.

The instruction set includes the ability to load any register with the effective address, i.e. the address which has been computed and which would have been used, is put into the register specified. This is particularly useful for position independent coding. The other group of instructions useful for this are the conditional branches relative to the program counter. The M6809 has a full set of branch instructions with a full length address. The lack of such a set is one of the defects of the DEC PDP-11. The M6809 also includes abbreviated addressing. This gives faster and more code efficient access to 256 bytes of store. This area can be located in places throughout the store as the actual address used is the concatenation of the lower eight bits from the instruction with the upper eight bits from the Direct-Page Register.

The supported address modes are immediate, direct, indexed and indexed-indirect. Indexing is supported with automatic pre-decrementing or post-decrementing. The modes are completed by base/index and program counter relative as already mentioned.

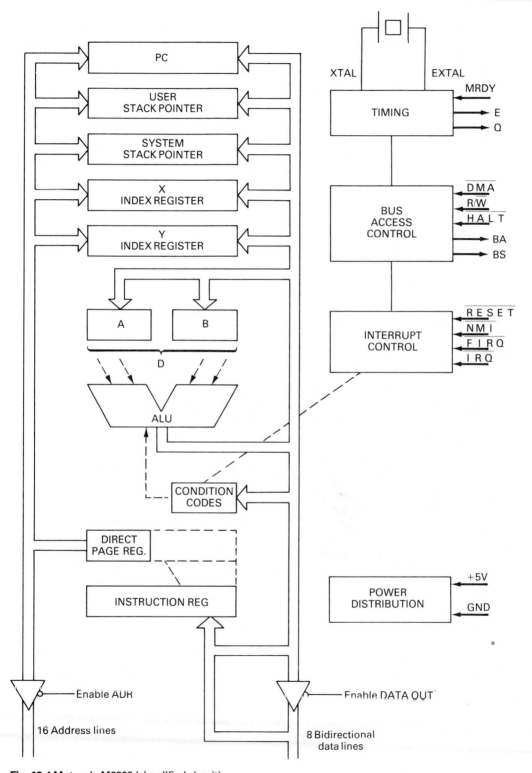

Fig. 13.4 Motorola M6809 (simplified circuit)

The processor is housed in a 40-pin Dual In Line (DIL) package with separate address (16) and data (8) paths. There are two power lines and two to output the processor status. This is either normal or an acknowledgement for interrupt, synchronization, halt or DMA grant. Two inputs connect to a crystal to control the internal oscillator speed, and two outputs provide clocking signals to the other chips. Powerup reset is complemented by three interrupt lines: normal, non-maskable and fast, which only stores the minimum of registers. Signals to handle direct store access contention and extending store access for slower store chips complete the pin out.

Software support for the 6809 is extensive. It is quite simple to write systems for it and OS9, a popular multitasking system taking less than 16k of store, is commercially supported. Similarly, Pascal and C compilers are available and produce efficient code for running on an eight-bit machine. The 6809 is arguably the best eight-bit micro. An earlier and more simple form is available as the 6800. Packaged with a 128 byte read-write store this becomes the 6802. When 2 kilobytes of mask programmable ROM, a single counter/timer, serial I/O interface and some parallel I/O lines are added, the device becomes a single chip micro-computer, the 6801. As such it is well suited to integrated interfaces and controller applications, though costing quite a bit more than the TMS1000.

13.4 Example 16/32-bit Processor: NS32032

The National Semiconductor 32000 range of processors and supporting co-processor chips represent the closest microprocessor designs yet to the mainstream minicomputers. They are even superior in some ways. There is a number of different processors to the same architecture with 8-, 16-, or 32-bit wide external data buses and various speed options.

They implement a pipelined, orthogonal architecture with a powerful set of instructions. There are co-processors to support demand-paged virtual store management and floating point operations. The processor chips contain all that is necessary to integrate these co-processors, for example instruction abort and restart for failed (paged-out) store accesses. Internally the processors are 32-bit machines. They have eight completely general-purpose registers 32 bits wide. Special purpose registers, curiously only 24 bits each, are supplied for the program counter, interrupt stack and user stack pointers, interrupt vector table base and two data base registers namely, static base and frame base. Two 16-bit registers hold the status and a pointer to the module descriptor of the current software module. Finally there is a configuration register to show the presence of co-processors and interrupt vector circuitry. Limiting the special registers to 24 bits constrains the architecture to 16 megabyte address spaces though, happily, addressing is completely linear.

There are about 100 basic instructions including full 32-bit multiply and divide and a variety of string operations. Some specific instruction types have been included to improve the efficiency of compiled code. Control of loops is simplified by the <*Add-Compare-and-Branch*> which performs the entire loop-end control in one instruction. The **Case** statement is supported by a multiway branch which indexes down a table of branch offsets. The need for local variables in high-level language procedures is supported by the stack frame allocation and the frame base register. The instruction set is orthogonal providing support for various lengths of its seven basic data types: bit, byte, bit field, boolean set, BCD, integers and floating point numbers. There are four basic addressing modes, register, immediate, absolute and indexed, and five modes to support high-level languages. These are base-indexed, store-relative, top of stack, scaled-index and external via a link table. Address displacements may be full addresses, or short forms (chosen by a two-bit field) to avoid the overhead when only a short offset is needed. Perhaps one of the most important features of the instruction set is that with only minor exceptions it is a true two-address machine. This means that $X:=X+Y$ can be specified in one instruction with the variables X and Y resident anywhere in store.

The NS32000 series arguably represents the best micro architecture to date, rivalling even the best minicomputers such as the Digital Equipment Corporation's VAX machines. It is very easy to write good assembler code for the 32000 series and with the extra support instructions code generation is made simple and efficient for compilers too. So chips can now be made with enough transistors to give an unrestricted architecture.

13.5 Example Transputer

There is, however, a completely different approach to microprocessor chip design which is to use the minimum architecture. With very large scale integration the actual circuit elements become even less important than in previous technologies. It is the inter-connection—the wiring—which dominates the design.

One way to break out of the strait jacket is to simplify the architecture, giving the reduced instruction set computer (RISC). A second way is to use repeated, regular structures. Stores are a very good example of this. Finally if a simple communication mechanism can be arranged one ought to be able to repeat processors over and over to build up more powerful systems. The INMOS Transputer represents just such an approach. The first released version of the family, the T414, is a 10 million instruction per second RISC design. Shown in Fig. 13.5, it includes on chip a 32-bit arithmetic and logic unit and a 32-bit, 25 Megabyte per second store interface to extra off-chip store. There are 2 kbytes of 50 ns store and four high-speed serial links which can operate at ten million bits per second on the chip. Other transputers offer different processors (e.g. 16-bit or with on-chip floating point hardware), on-chip store and I/O capabilities. The T414 contains about 200 000 transistors.

The serial links are a particular feature as it is their presence which allows the Transputer to be considered as a 'component'. They drive at TTL levels and are run isochronously. The transmission clocks are generated by doubling each Transputer's main clock, so any phase relationship but only very small frequency differences are tolerated. Because of the speed there are strict limits on pulse spreading, and rise and fall times, so the ports can only drive about 40 cm of PCB track directly. Longer distances require additional circuitry, simple 50 ohm series matching giving 3 metres or so.

Apart from the impressive speed the protocols and higher level operation are a good step in the right direction. The isochronous data format has a start bit $<1>$ and a type bit. If this is $<0>$ the message is an acknowledge message and that is it. If the type bit is $<1>$ it is followed by eight data bits and a stop bit $<0>$. This acknowledgement gives synchronization at the byte level. A link between two Transputers is two data wires, one in each direction, with data frames and acknowledge frames multiplexed onto them. Each acknowledge indicates that a data byte has been received and that another may be sent. Hence continuous, bidirectional, DSA transmission can exchange 1.5 M bytes per second on each of the four links. The processor's performance drops to about 80% because of the loss of store cycles to the channels when all four run flat out.

In another reasoned break with tradition no machine code instruction set was released and supported for the Transputer. A higher-level language **occam** is the lowest level for users. Hence better transputers can be released without requiring any change to the software, or being caught by 'upwards compatibility'. Based on the CSP work of Tony

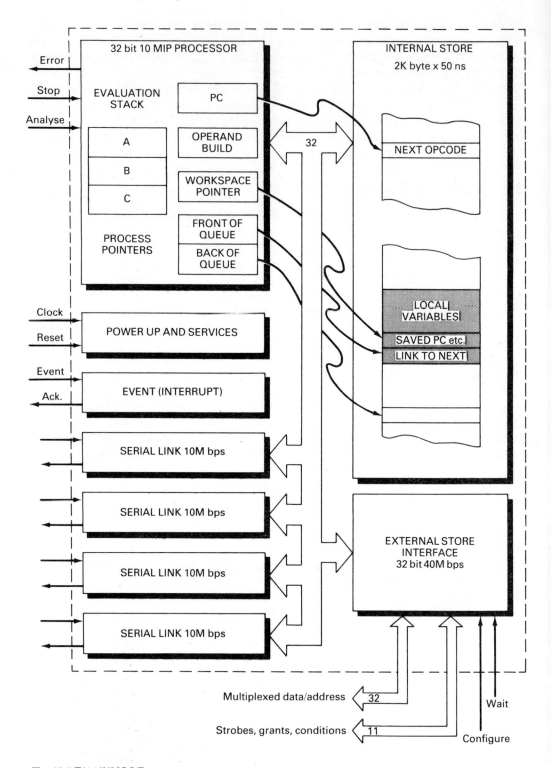

Fig. 13.5 T414 INMOS Transputer

Hoare, occam is designed to exploit concurrency from the outset. A program, written as a set of communicating sequential processes, can run on a single Transputer or, in an identical fashion, on multiple Transputers communicating via the links. To make it fast on a single processor, process scheduling is done in hardware and with the small register set context switching takes a microsecond on average, two and a half in the worst case. The processes connect by channels which are the links described above or store locations within one Transputer. At this level they operate as fully synchronized, zero-length 'pipes', or to use a decade-old term 'flanges'. Process priority is either:

(a) to run to completion or until it must wait for channel communication, or
(b) queue for a time slot, which is used until communication is needed.

All in all the Transputer represents a significant new approach to microprocessor application. They will have extra on-chip hardware support for speed critical applications, such as signal processing and graphical output, when it is needed. Otherwise they can be used as general-purpose programmable components. Support chips are available to convert the serial links to parallel I/O much like the chips described in Chapter 12. When compatible chips come with all the analog circuitry to interface directly to the serial links the Transputer's claim to be a processing component will be justified.

This chapter has shown some of the problems of choosing a suitable processor on which to base an application. General choice criteria and specific points for architectural comparison show us what we should choose, but as often as not company history or executive direction will cloud the issue.

14
Design Techniques and Support Systems

Perhaps the two most limiting factors affecting interfaces in the past have been the absence of rigorous design techniques and the lack of adequate support systems to permit the designs to be produced speedily. Rigorous design techniques have existed for logic design from gate level to some medium-scale parts for a long time but these have not been applicable to programmable and large scale integrated circuits. Software design was originally a very ad hoc process and it is only relatively recently that more firmly based design methodologies have become common. Sadly a lot of interfaces are still supported by programs written in the 'hacker's Basic' style. The process of integration of separately designed software and hardware, and to a greater extent the testing of systems (and programs), has received little attention. The vast majority of programs are never properly tested prior to integration, and following this, further work is only done when 'bugs' become apparent. For the control of any complex real-time system such an approach is a sure recipe for serious disasters.

14.1 The Design Process

Any project of this type divides into hardware and software parts. Each then proceeds through set design stages to an iterative 'test and correct' cycle and finally to an iterative integration phase of the two halves. This process is outlined in Fig. 14.1. This is a reasonable way of organizing a project and the design phases shown can be rigorously controlled but the software side has always been problematic due to the inadequacy of testing and in many cases the lack of rigour in design. The integration phase has often been the downfall of entire projects causing incessant delays as seemingly trivial faults are searched for. This is, again, usually caused by the inadequacy of the test environment and the lack of a correct overall specification. The hardware side causes fewer problems because it is often less complex and rigorous design techniques with design-rule checking have been standard practice for some time.

Turning to the software development process, a typical cycle for small projects or modules of larger projects is shown in Fig. 14.2. Good tools exist for most stages: editors, compilers and optimizers, source code and object library maintenance, etc., in fact for all stages up to testing. It is then when the programmer is left on his/her own.

As has been observed many times before:

Testing only shows the presence of faults, not their absence!

The usual technique for testing is to run the program or module and observe the results. Various test data or test harness arrangements would be produced by the programmer to assist in trying to expose faults in his design.

Running a program in its **real environment** to test it can be likened to trying to learn to fly an aeroplane by being passed the controls at ten thousand feet and being told to land it by the previous pilot, who then bails out and leaves you to it. After the (almost) inevitable

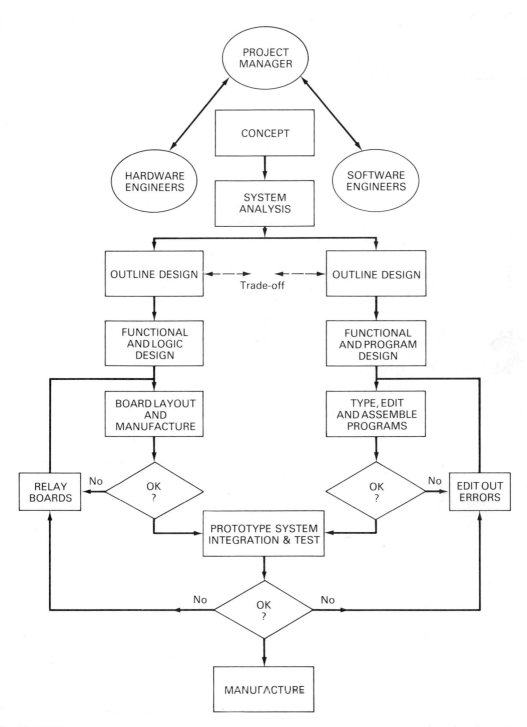

Fig. 14.1 The overall design process

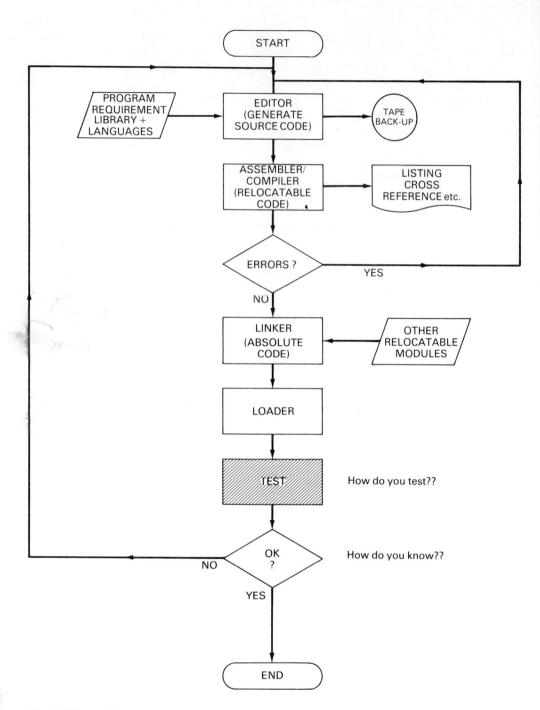

Fig. 14.2 The traditional software design process

crash the débris is examined and conclusions drawn. There may be assistance from a 'black box' recorder or its equivalent a store dump!

As an improvement on this a **simulated** environment is sometimes created which allows the program to be run, albeit at much reduced speed and incapable of real device input/output control. The ability to trace its operation and monitor states and variables is included. This is obviously better than uncontrolled running to test, but an even better method exists which will be discussed in greater detail further on.

One can liken the simulation test approach to learning to fly in a flight simulator. It is good—one can even be airsick, but there is always something missing—it is **not** real and does not exactly simulate reality so there may be cases when it is incorrect and these may only show up in real usage with possibly disastrous consequences.

The way much software is developed has always been mystifying to the author. In particular, three aspects of the way most people develop software stand out. For example:

Why do programmers insist on:

(1) being so obscure,
(2) regularly throwing away half their work, and
(3) continually reinventing the wheel?

There are simple solutions to all three but, as they are not in general use, some support is necessary for the statements above.

The **first** can be summarized by the game played between users and designers called 'guess the command'. The designers or programmers of command interpreters, editors etc. carefully choose their input formats and commands (e.g. delete, erase, purge, rm, remove, etc., to expunge an object or file). They program their syntax checker to ensure that only legal commands will be accepted, or suitable error messages given. The user then searches the documentation to find the command format to use at any point!

Of course as only a few command words will be accepted it seems quite ridiculous to let the user even enter anything else, only to reject it.

The solution is to adopt uniform **directed-syntax** control for all input, **not** to be confused with simple menu-driven systems where you 'type 1 for edit' etc. A set of keys is defined (continually redefined) to represent the legal inputs at the time—an alternative is to use a 'mouse' to select keywords on the screen, perhaps from pull-down menus. A single key depression then enters the word and the syntax is stepped to offer the next legal choices almost immediately.

This is obviously much more efficient and convenient for the user but it is also easier for the designer. If only legal syntax is entered the checker and error handler are much simpler! Documentation is only rarely consulted.

The **second** mystification is that after going to all the trouble to produce elegant programs with meaningful names for identifiers, compilers producing runnable code ready for testing carefully throw all this away. Control and testing is far harder as there are few links to the original code. It is common to suffer 'error C at 5921', 'too much input at line 2100', etc! The solution is to **retain** the compiler or assembler **symbol tables** and keep them through any linkage phase so as to be available during testing. This allows direct ties back to the source program without the hopeless inefficiency of run-time interpreters like BASIC.

The **third** aspect which also makes testing unnecessarily difficult is the reinvention of the wheel. It is surely not reasonable to have to write yet another sort routine for each new project. Furthermore if we accept that 80% of the design takes 20% of the effort, and the remaining 20% of the design takes 80% of the effort, one can make the difficult 20%

reusable to simplify subsequent projects. The reason for the difficulty is often that this part is highly time-critical or uses much detailed input/output. A technique for producing, and using, standard or canned software is needed. After all, one may have chosen a specific microprocessor so that the same hardware can be used as in the previous project. It is a small step to extend the idea to many parts of the code.

As an aside it is also beneficial to use an architecture which supports re-entrant, or pure, code. This gives us the ability to invoke a procedure when it has already been invoked by another or the same procedure, with a new set of parameters. It is particularly important for multi-tasking systems as a single routine may service many tasks. Each new task then only needs extra data space—not extra program space for the shared routine. A consequence of needing re-entrant code is that the architecture must support data manipulation on a stack, must support register saving or switching in indivisible operations on interrupt or context switch, and must perform address manipulations indivisibly.

One should also note that real-time systems with lots of **new** hardware, software and users are commonly complex in structure and so testing should start with small units in isolation. A central reference specification is essential as the basis for the initial design.

14.2 Canned Coding Techniques (CCT)

There is a number of problems which have mitigated against the adoption of CCT in the past. If a piece of code is to be used again (be canned) then it must be capable of re-use without alteration as any change is likely to introduce errors to an otherwise well tested module. This means that the computer architecture must support position independent coding. Secondly the module must be linked to other modules by a parameter passing technique which is simple to understand, efficient in use and supports high-level language programming. Thirdly the module must be worthy of re-use: i.e. greater effort than normal must be put into selecting the right algorithms and in the quality and accuracy of coding them. The management of programmers to do this is beyond the scope of this book, but the support architecture to produce code in read-only store modules which can simply be plugged together to work, is not. Until recently there were few architectures to support CCT fully. The Z80 (and all Zilog processors to date), the 8088 (and all Intel processors to date) and most mainframe computers are deficient in this area. The requirements are not that stringent, though:

1   **Uniform address space**
2   **Relative addressing of data (constants and variables)**
3   **Relative addressing for control transfers**
4   **Stacks with full stack manipulation**
5   **Indirect input/output (easiest by store mapping)**
6   **Transparent interrupt handling and context switching**

As a canned module may be used at any location, without alteration, there must be no restriction on addressing. This implies a uniform address space which may perfectly well be paged or segmented on arbitrary boundaries. It must not have any artificial limitations such as fixed-size segments. As the canned module may be located at any address it can have no *a priori* knowledge of any absolute locations. Its local constants, stored in the same ROM, must be addressed relative to the program counter. Its local variables will be located in RAM, the location of which will only be available at run-time. Thus all these must be addressed relative to some base register other than the program counter, or be placed on a stack. The jump (JMP) and jump to subroutine (JSR) instructions are of no use as they have absolute addresses fixed at compile (link) time. All transfers of control must use branch instructions. These can be conditional or branches to subroutines. Many machines, such as the Digital Equipment Corporation PDP-11, provide these instructions but assume they will be local and so have a reduced addressing range. To use these cut-down forms for longer transfers of control implies a chain of a branch to a branch etc. to get far enough. The branch instruction set must provide the full addressing range of the machine. It may give abbreviated addresses as an extra option for the compiler to save space and speed where they fit.

When we come to communication between separate canned modules there is a number of choices. A fixed common area of store could be set side. This has all the problems of FORTRAN Common: it is unstructured and undesirable. We could pass all parameters through registers. This is restrictive in the number which can be passed unless the number of registers is increased, or an overlapping set shared between two modules is implemented. We could pass addresses of parameter areas via registers. This relies on the communicating modules knowing each others parameter variable structure and relative locations. Finally we have the best choice—use a stack. Once high-level language designers realized the benefits of stack operation their use became commonplace. For CCT such use is essential. The calling module places its parameters on the stack and can reserve the space for the return parameters. When passing a variable number of parameters one would include a count of the number and type as the top parameter. The called module can use the stack for its local variables, and any similar further nested calls, with impunity. It finally fills in the parameters it is to return in the space provided and returns. This mechanism means that indexed addressing relative to the stack top must be available for all data-using instructions.

Perhaps the most complicated part of canned modules is handling I/O—yet again! The I/O addresses are bound to be absolute, but it must be assumed that they will be different for each system using the canned module. They will have to be passed into each routine needing them as parameters via the stack. They will have to come from somewhere though. We will also have to have some way of tying together the canned modules from separate ROMs which do not know of each other's location. Interrupt handlers will not be at fixed locations so the solutions described in Chapter 11 will be needed. Transparent, indirect vectoring is implied, and the use of workspace switching to change context quickly is highly desirable.

Tying all this together we can see that after plugging in RAM and I/O chips we add our 'standard' ROMs and need one extra. This **configurator** ROM is used to define the absolute locations of all the other chips and has small routines to set up the initial parameters and vectors etc. It contains the basic powerup code and so may be located to include the reset vector directly. Figure 14.3 shows the accesses we are trying to achieve, and the arrangement of chips with the configurator. One may obviously take a number of separate CCT modules and merge their contents, blowing them into a single, larger read-only store chip.

14.3 Finite State Machine Design

To ensure the quality of the re-usable modules it is as well to have firm formal specifications for an entire system. Traditionally, specification of software is done in a programming language and hardware by logic, circuit and timing diagrams. To be more unified and precise we can use the Finite State Machine (FSM) approach. The FSM applies equally to hardware and software and consists of a series of **states** and a series of **transitions** between them. There is much in the literature on this and the related, but more comprehensive, **Petri Net**

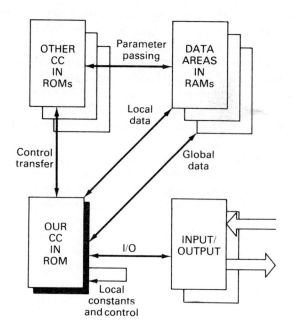

(a) **What we want**

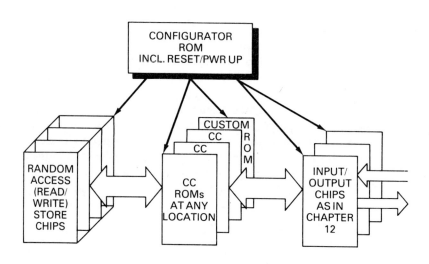

(b) **How we get it**

Fig. 14.3 Canned coding techniques

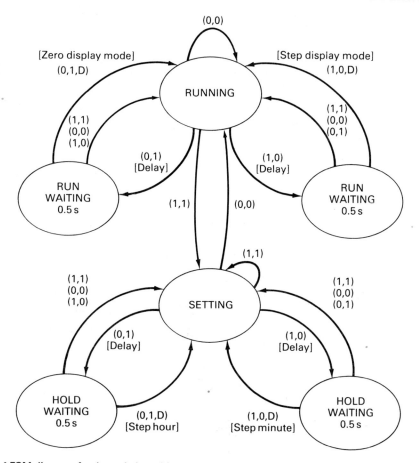

Fig. 14.4 FSM diagram for the switch problem

diagrams, which explicitly include both states and events (actions and transitions). States are shown in an FSM diagram by circles, and transitions by a directed line. In the example shown in Fig. 14.4 two switches are to be used to set a clock. Depressing both switches resets the clock to midnight. From this condition we may then release one switch or the other. One causes advancing of the hours hand, the other advancing of the minutes hand. Once the correct time has been set both switches are released and the clock runs. We may then use one switch **or** the other to cause other actions, for example selecting different display modes for the time. Pressing both would again reset the clock. The switches are assumed to be debounced either by hardware or by software action, say only checking them every half second and insisting they remain in a given state for half a second before being taken as an input.

The FSM moves from state to state as the result of a stimulus, in this case a switch changing, or a delay ending. Each state change is also associated with an action, though this may be null. It is quite simple to convert from a complete FSM diagram into either software or hardware. It is also a simple matter to draw up a transition table which shows for each state the next state(s) and the stimulus necessary to reach that state and the action to be performed as a result of the change. Systematic evaluation of the table shows if there are redundant states which cannot be reached, duplicate states or deadlock states from which there is no exit. We can even automate this directly into a program or hardware. For a program we simply require a data structure to specify the table and for every event during a state give a procedure to perform the required action and the identification of the new state. The FSM and its transition table forces the person specifying it to be explicit about all possibilities, leaving nothing to the chance of coding.

14.4 The Controlled Test Environment

Manufacturers provide various forms of test environment, particularly for testing embedded microprocessor-based systems. **Simulation** on a processor other than the target gives no information on how well the target hardware will integrate with the software. The simulation can also be subtly misleading as the simulation is never exactly correct. A simulation can be run on the correct target processor but with standard hardware for the rest of the system. The program logic will be tested correctly but integration will again be untested until the real target system exists.

Once prototype hardware is under way one can start running the programs for integration. The problems of running for real have already been seen so we use another approach—**emulation**, and in particular In Circuit Emulation. Here the target processor is replaced by an identical one in a 'cocoon'. This allows full control and with the addition of suitable monitoring hardware allows us to observe the program and hardware running together **exactly** as they will in the final system. If anything goes wrong then as well as a trace of what happened **before** it went wrong we can prevent the disaster of crash landing! We exercise the control option and stop the offending portion of program. We can then switch to a tested safe clean-up routine to keep the external hardware happy if necessary. This is similar to learning to fly with an experienced pilot, dual controls and a black box recorder! We can **trace** and **breakpoint** programs in full real-time on full real hardware. No other approach offers this.

Figure 14.5 shows the organization of an emulator system and the three major parts: the processor emulator (for control), the store emulator (for the program) and the logic analyser (for observation). Any one of these is an improvement on a simulator or real-running unaided. The combination of all three is an unbeatable tool. Important points to note are the separation of the emulation/target store from the host processor's store. This places no restrictions on the mappings we may perform on the combined emulation/target store. Similarly separation of the host processor's bus from the emulation bus gives us operation without any speed restrictions.

14.4.1 Store emulation

We are to run programs on real prototype hardware. The final programs will commonly reside in read-only store. During program testing and integration the cycle of edit, compile, link and run will be repeated as in Fig. 14.2. We can considerably shorten this cycle by the techniques already seen but also by getting the program into its store quickly. Programming an EPROM (erasable programmable read-only memory) chip or chips each time may well take longer than the rest of the cycle. Down-loading from a host at terminal-line speeds is similarly slow. Also once fixed in an EPROM we cannot try parameter changes, small program fixes, etc., without going round the whole cycle.

The store emulation consists of a controller, an address mapper and a reasonable amount

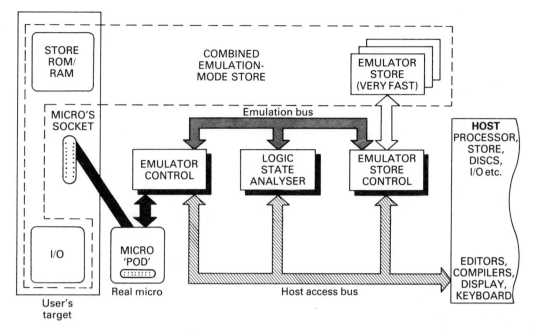

Fig. 14.5 Emulator system components

of the fastest static store available. This store can be allocated in smallish blocks (say 1K) to any addresses in the store map. The mapper allows us to mix emulation store with real store and store mapped input/output devices on the target board. The distance between the real store on the target and the emulated store may be a few feet which, with the added mapper logic for signals to pass through is the reason for using the fastest available store chips. The net result is a composite store running as fast as will the final real store. It can be loaded from the host very quickly. It can be modified during testing by host command. We can also use the controller to signal if we misuse the store by addressing where there is none or writing to store which is pretending to be read only. With the separate emulation and host buses we can also take sneaky looks at the emulation store **while** the program is running at full speed. The store emulation is a useful tool on its own.

If a processor emulator, as described in the next section, does not exist for the processor we wish to use then we can still achieve say 80% of the test power. Using the emulator store to emulate ROM or EPROM will allow us to load our programs quickly and analyse all store accesses. A separate means will be needed to stop and start running, but at least we can monitor our program running in real-time. This is called ROM emulation and is a good half-way house.

14.4.2 Processor emulation

To give full control the processor chip is unplugged from its target socket and a lead is plugged in which will behave in an identical manner. This can be achieved by using a more powerful machine 'B' to emulate the target processor 'A'. This is expensive and still may not be perfect. The perfect emulator processor does exist. The system emulates 'A' by itself, another 'A'—again the best available. This processor is then surrounded by switching logic

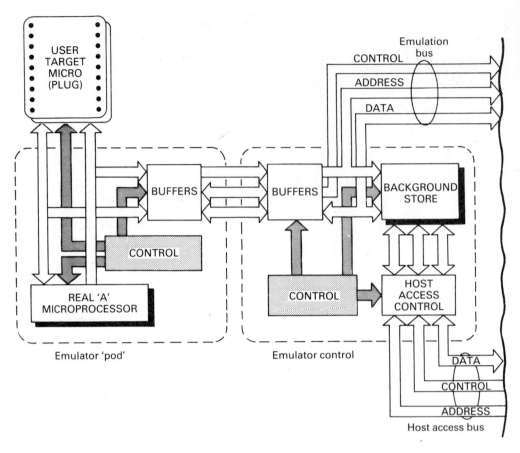

Fig. 14.6 Processor emulation

to allow it to be connected to the target when running the program or to the host emulator control when it is to be stopped and register content observed etc. Remember micros don't have a front panel of lamps and switches to see the register contents. The host connection is used to run special programs (from a separate hidden or background store) to load and change target store or I/O and to view their contents and register contents. All of this extra logic must be transparent though.

The detailed design of a processor emulator is shown in Fig. 14.6. Important features are the background store with space for communication data and the special program, and the control buffers to switch the processor between target and host. A good emulator will meet all the requirements of Table 14.1 with the possible exception of the electrical conditions. Emulator buffer drivers are usually **too** good, being more powerful than those common in micros. Also the need for a length of ribbon cable from the plug to the pod (which contains the actual processor) increases the observed pin capacitance of the 'micro' somewhat. Caution must also be observed with some microprocessors which rely on obscure secondary effects with which emulators may have problems. As an example, some manufacturers' integrated I/O chips rely on seeing the Return from Interrupt (RETI) function code on the bus to clear the device's interrupt request flag. Depending on the buffer arrangement they

Table 14.1 Processor emulation requirements

•	Functional Transparency	instruction execution
		store accessing and DSA
		stimulus-response, interrupts, co-processors
•	Temporal Transparency	clock speed—real-time
		bus activity
		store access—no wait states
•	Electrical Transparency	loading and capacitance (fan-in)
		drive capability (fan-out)
		propagation delay

may not see it as the only transfers expected would be the address bits from the processor to the store module and the op-code from the store back to the processor.

14.4.3 The logic analyser

The key component for advanced testing of digital systems is the part which shows us what has happened. When digital logic systems were simpler than today a dual-trace oscilloscope sufficed, but it shows what **was** happening not what **had** happened. An oscilloscope only displays traces after some triggering event and continues to display them only so long as the signals repeat. The addition of an expensive **storage tube** is needed to permit an isolated period of events to be captured for later observation. For complex microprocessor-based systems, and those incorporating other programmable devices, a more extensive device is required. The oscilloscope is inadequate as these micro-based systems have few, if any, repetitive signal patterns from which it could trigger to display a static picture. The name now given to the class of extensive tracing devices used is the **logic analyser** and they have the following features in common:

1 **Data is stored prior to display**
2 **Data is captured on a time or event base**
3 **Data to be captured may be restricted to certain types of events**
4 **Commencement or termination of storage occurs on arbitrarily complex trigger patterns or sequences**
5 **Data may be post-processed between capture and display.**

The oscilloscope still has a role for the parametric measurement of a specific signal, often being triggered from the analyser.

Analysers, such as the one shown in Fig. 14.7, have three types of signal input. The first are the **main** signals which are to be monitored. In the majority of our cases these will be the address and data buses and various control lines from the microprocessor. The threshold of these signals is usually fixed for the particular type of logic being monitored (TTL, CMOS) but it can be variable. It is also possible to have two threshold points (but we need double the store) to show the rise and fall times of signals.

Clock inputs come from the target system for **state** analysis. An internal free-running clock oscillator provides the timebase for timing analysis and glitch detection though. Finally clock **qualifier** signals are the same or additional control lines used to pre-process the main signals prior to storage. All signals can be selected with either polarity or as 'don't care' conditions (X). The qualifiers allow us to select **what** data to store, such as just store reads, just interrupt acknowledgements or just operation code fetches.

A separate section of logic determines **when** we store the data. The store is arranged as a cyclic buffer operating on a first-in first-out basis. The clocked, qualified data is continually stored and we simply need to say when to stop storing more data. Three times are commonly useful. We either stop when we reach the **trigger point** thus giving a store of events **before** the point of interest, or we stop a full store sweep **after** the trigger point, or we stop a half store sweep after the point giving a trace **about** it. The last one is most useful as it shows what led up to a point of interest and the events following it.

A problem is now apparent which will affect our interpretation of the data we will see on

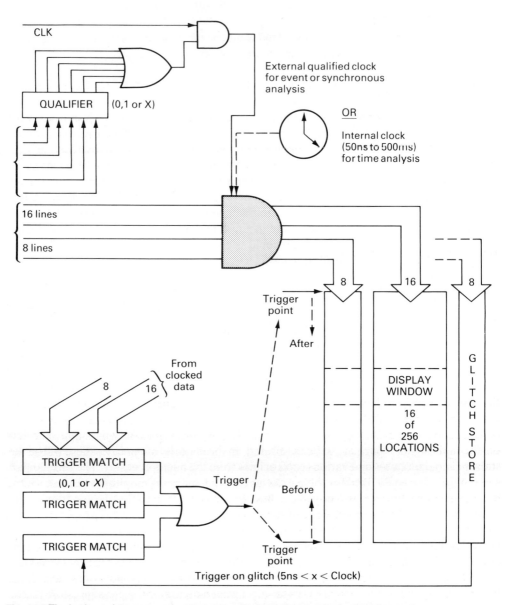

Fig. 14.7 The logic analyser

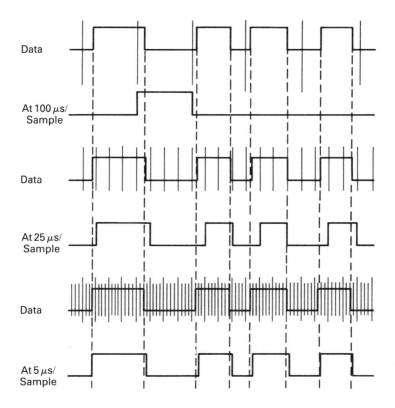

Fig. 14.8 Effect of logic analyser time/state analysis

the analyser's display. We may be storing information with a fixed **time** base or on a qualified **event** base. In either case data is only stored on the qualified clock edge. Figure 14.8 attempts to show the effect of different time base resolutions on the data we observe. One must always take care to ensure that what is on the display is not confusing. For example a regular clock at twice the speed of the sampling clock and fractionally delayed behind it will be observed as a steady logic zero! In many analysers a separate store is included to help avoid this confusion. For the purpose of this display we extend the definition of a glitch from that in Chapter 10 to include any signal which crosses a threshold point and does not reach the opposite state, or remain in it, until the next sample clock. They (the super glitches) can be displayed by a single bright line on the logic trace or by brightening up an edge on which one occurred. A comparison of real and observed signals is shown in Fig. 14.9 and the glitch store hardware can be seen in the analyser of Fig. 14.7. The user simply has to increase the sample clock rate to show whether the glitch is genuine or is just a pulse or signal mis-observed because of slow sampling.

We are, however, back with the problem from Chapter 3 of what is the flow of time, what is a point of time etc.? This leads us into research areas of temporal logics, but it is worth noting that all time for our systems is relative. Any measurements are made over an interval and any evaluations are at one point with reference to another point in time. Also we must guard against the temptation to read more into the data than is really there. It is impossible to capture all the available data so we restrict what we store. We end up with a nice display of very limited data.

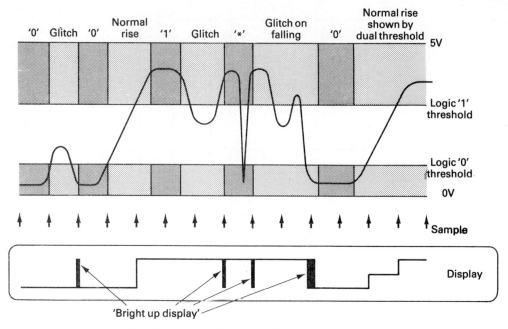

'*' Legitimate signal but too fast for sample rate ≡ glitch

Fig. 14.9 Glitch detection and display

The trigger logic is quite simple though. A number of registers are used to hold patterns (0,1 or X) which describe the point(s) of interest. All the input signals and qualifiers are represented. A simple set of exclusive-OR gates compares the stored patterns with the incoming clocked data. When a match is found the trigger signal can be used to halt storing data as described or to activate another of the register/comparator pairs.   .

So we could put together the analyser with retained symbol tables, directed syntax and emulator processor control to enter the command:

trace about address = *Sort_routine* **status** = **opcode occurs twice then status** = **intrptack traceonly   address   range** = *Sorted_list*   **through**   *End_list*   **data** = 1XXXXXXX **status** = **write haltprocessor on tracecomplete**

in only 60 key strokes (instead of over 200!). It is quite an easily understood command! Directed-syntax words, for which only one soft-key depression is needed, are shown as bold. The fault we are looking for causes our real-time controller to go haywire after we call the sort routine the second time and when we try to sort negative numbers after an interrupt has been serviced!! You can see the effect of triggers: three trigger points are used, and the qualifiers to limit the data put into the store to that we are interested in. We also have full control so the processor can be stopped and we can then see all other store states and register contents. We could switch to a trusted program to ensure the safety of external hardware immediately our fault is detected.

Programs can be tested in small parts as we wanted, but more important is our ability to post-process the stored data. Had we traced instructions, or all states with the command above then we would have had a lot of really meaningful zeros and ones to look at! By having the host processor run a program to translate the patterns back into instruction mnemonics we can see the trace more clearly. Remember the trace data is stored with its status e.g.

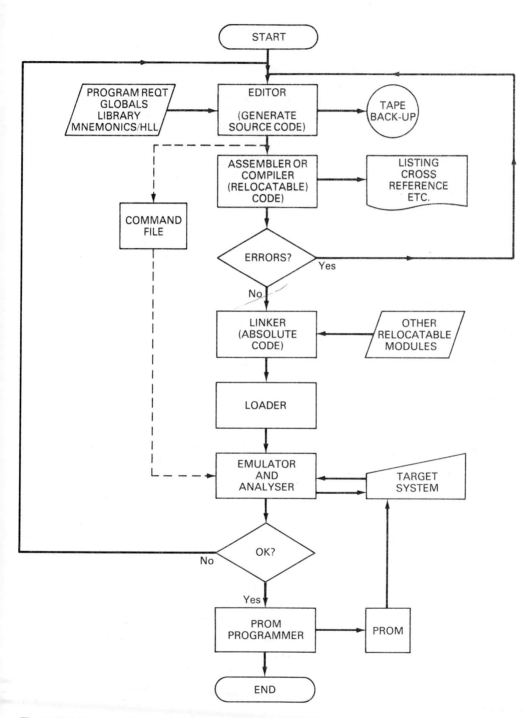

Fig. 14.10 Advanced software design and testing

opcode etc., so the disassembly program can be very detailed. Giving it the saved symbol tables lets it put back our meaningful names and tie the trace back into our source code before displaying it to us. This takes only a second or two of host processor time.

14.5 The Microprocessor Development Unit

So we come to the complete development process we would like. Figure 14.10 shows the changes needed from the original attempt of Fig. 14.2. It also includes the integration phase as the use of linked emulation and analysis tests the combined hardware/software target system. Apart from the benefits already discussed, speed of operation is the major advantage of this approach. When the author first used host machine facilities to develop microsystems in the early 'seventies, the time from one test finding a fault through rectification to retesting took over two hours. Program editing was a slow, crude, line-by-line job. Program assembly, also run on the computer centre's machine, was slow because of batch program size restrictions. Object code then had to be loaded into the target using 30 character per second terminal lines which were the fastest the centre then supported. Before logic analysers were available, testing was haphazard to say the least, and the only amazing thing is that our systems ever worked.

Extensive studies and experiments carried out at Yorktown Heights by International Business Machines Inc. have demonstrated conclusively that shorter response times make users more productive. This seems quite intuitive, but the actual figures are most important. There is a good but linear improvement in transactions per hour as the system response time is reduced until it reaches about half a second. Once the response time drops to about a third of a second or less there is a dramatic improvement in transactions per hour, particularly for expert and average users. Even novices demonstrate a noticeable increase.

To achieve fast development times and high productivity, we employ all the technology we can, particularly that already discussed. A high-speed workstation including the host processor, store, hard-disc backing store, emulator and analysis system must be devoted to a single user for development and testing periods. Such a system is expensive, the more so if it is needed to support many different processor types as in a complex organization. Also, if such a system is used, development and testing times are reduced so much that the system should be idle for much of the time! Consequently wherever practical co-operative units are established to share the expense over a number of development groups. As an example, the Wolfson Microprocessor Research Unit was established at Imperial College in 1980 with the aid of a generous grant from the Wolfson Foundation. Since then the unit has supported dozens of projects by providing the equipment and expertise that has been described. The projects have ranged from a hierarchic, multiprocessor control system for an industrial robot arm and an automatic guided vehicle, through low-cost expert systems, local area networks and a wind-driven turbine controller, to high-speed stepper motor, multi-language typewriter and quality printer drivers. The only common factors were that all needed microprocessors (one had 10!), all were finished remarkably quickly and a fair number would have been impossible without this approach. The loop time from one test finding a flaw in coding, through rectification to retesting has been cut from a couple of hours to a couple of minutes and goes a long way to explain the successes.

15

Design Examples

Three fairly straightforward examples of interface systems have been included to highlight particular points. The display panel is used to show how very large numbers of outputs can be handled and to show how random logic can be replaced by a microprocessor. This book has concerned itself with real-world interfacing and an industrial plant would be a suitable design example. Unfortunately, it would also take most of a book to explain the plant, particularly a complex chemical one, before the interfacing and control problems could be discussed. A greenhouse has been substituted for the individual process as it is hoped the reader will already understand the general problems of its *environmental* and *plant* control.

The low-cost greenhouse system uses only one microprocessor as both data gatherer and controller, combining the micro as both the interface itself and the interface to the micro of Chapter 1. A completely portable system has also been included, particularly to show the use of high-level languages in interfacing and control. AURAC is a small mobile automaton which can be programmed to sense its own environment and move within its constraints.

15.1 Gas Plasma Display

The cathode ray tube is the basis for the majority of computer displays, but there are many circumstances when it is unsuitable due to size or lack of ruggedness. Small displays and annunciators may be built up in a number of patterns using **liquid crystal**, **light-emitting diode** or **gas plasma** display media. The character patterns can be formed in various ways, of which the most common are shown in Fig. 15.1. The best formed characters of these simple cases are produced in the 5×7 dot matrix, and the cheapest by seven segments as in calculators. Matrices with more dots are now also commonly found.

The display in this example has four rows of 32 characters, each composed of a 5×7 matrix of dots and an underline row of 5×1 dots per character. There are thus 5120 individual points which must be either on (*lit*) or off (*dark*) to give the display pattern. Obviously individual outputs for each of the dots would be impractical, requiring 200–250 PIA chips. The scheme devised and shown in Fig. 15.2 uses only two! This display is typical of a large range of small dot-formed panels.

(a) 7-segment   (b) 14-segment   (c) 5 x 7 Dot Matrix

Fig. 15.1 Segment and matrix display patterns

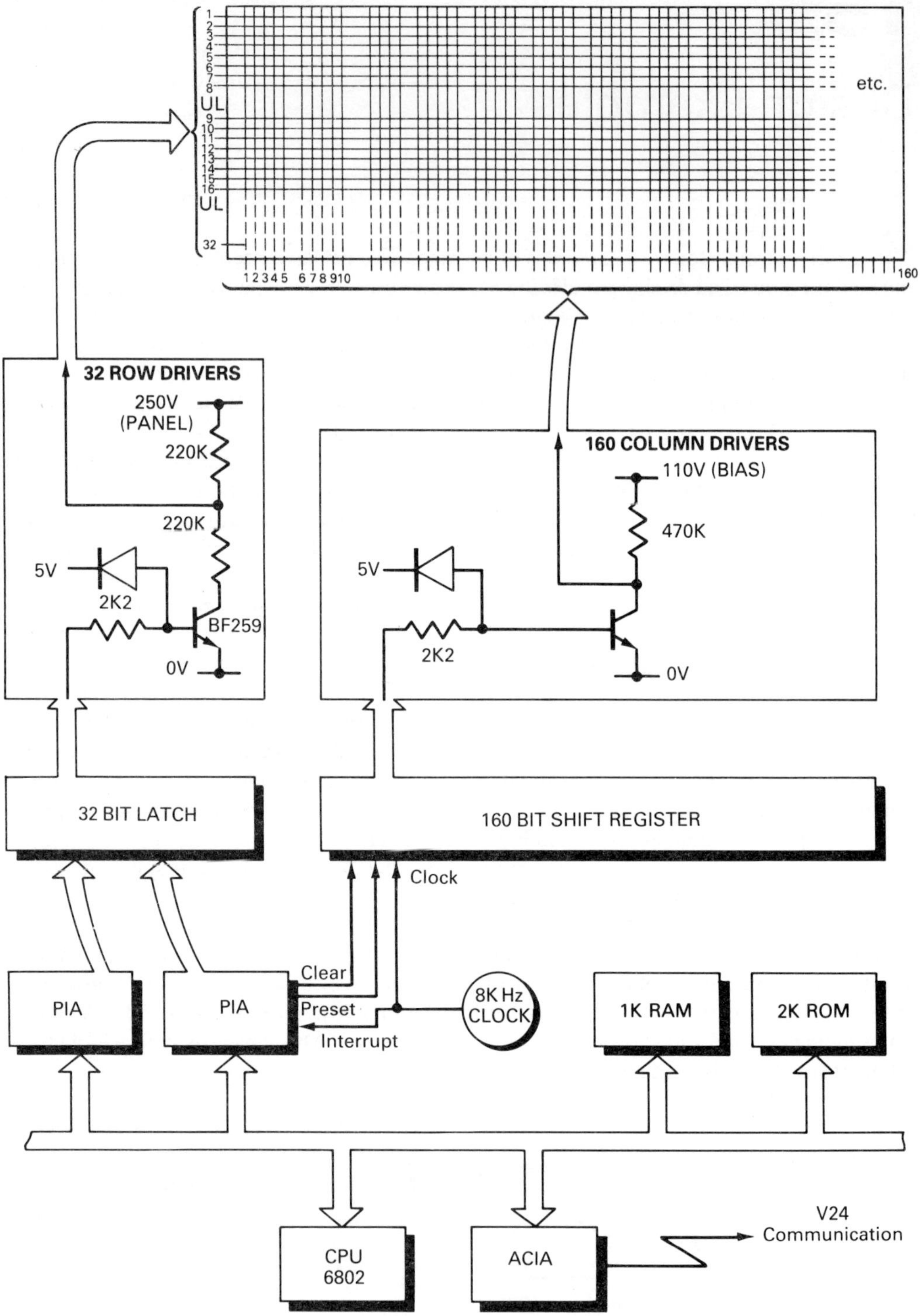

Fig. 15.2 Micro interfaced gas plasma display panel

Firstly savings can be made by recognizing that the design of the panel is important. Any light-emitting element will have two (or more) connections and both connections (ends) can be switched separately so that only when both are switched on will the element be on. The electrodes (anodes and cathodes) of each plasma cell are thus connected up in a matrix of 32 rows by 160 columns. Any cell will only light if the gas is ionized by applying a sufficient voltage. If half is applied to a row and half to a column only the intersection cell will light, others remaining dark. This cuts the number of interface lines to 192. A similar 'coincident selection' approach is used in all store chips. This arrangement also cuts the voltage switched by the drivers.

The second saving is made by time-multiplexing the columns. As only a single column can be accessed at an instant if the coincident selection is to give all possible patterns, the display is refreshed running along all the columns outputting the pattern for the row corresponding to each. They are refreshed fast enough to ensure a static display to the human eye. It is rather like a circus plate spinner who goes around continually topping up angular momentum to keep dozens spinning. This reduces the number of lines to the interface to 35, or only two PIAs.

The micro is the interface in this case, replacing random logic, as it does not decide on the patterns, it only reproduces those selected from its ROM by receiving the characters via the ACIA communication interface.

Overall operation consists of three routines. One, driven by a high-priority interrupt (e.g. NMI) caused by the 8 kHz refresh clock, extracts bit patterns from the display buffer in store (a row pattern of 32 bits per interrupt) and outputs them to the PIAs. The patterns are synchronously latched on into the latch drivers by the next clock edge following the interrupt. The second routine takes characters from the communications interface and stores them in ASCII in a large buffer in random access store. This routine is also interrupt activated but at a lower priority (e.g. IRQ). The final program runs continuously and scans the communication buffer for display commands etc. and produces the interleaved bit patterns in the display buffer for the currently displayed part of the communication buffer using stored patterns from the ROM.

15.2 Greenhouse

The microprocessor-controlled greenhouse presents an interesting design challenge, to control the environment it offers reliably using only a very simple (home) microcomputer. The control algorithms, display and keyboard are only briefly discussed as the problem lies in the interface and the parameters for the environment. Any commercial microcomputer would provide the human interface and programming capabilities anyway.

Considering the facilities the greenhouse shown in Fig. 15.3 should offer it is apparent that more than one environment is required. In this example two will be provided, one for seed propagation and one for growing plants in a soil bed.

The **propagation bed** is enclosed and of limited size, requiring only a small soil heater to maintain its temperature. As the air of the rest of the greenhouse is not to be permitted to drop below 5°C only about 60 watts of soil heating will be needed to maintain the correct temperature range of 18 to 21°C for the propagation area. It does need high relative humidity, however (approximately 80%), and so the enclosure will have a mist spray water inlet from the main water tank, controlled by a solenoid valve.

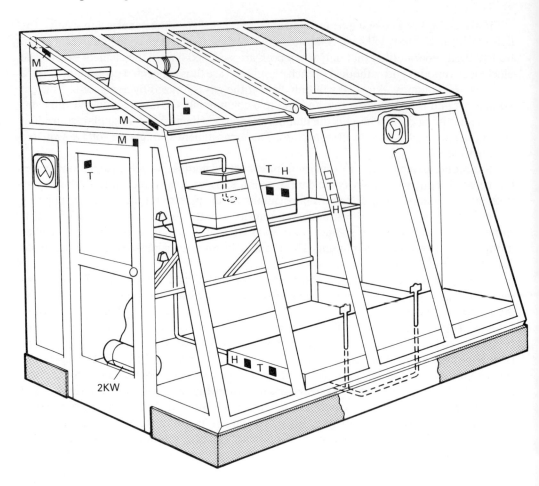

Fig. 15.3 The microprocessor controlled greenhouse

The main growing area has more points to consider, including:

- **Temperature control**
- **Humidity control**
- **Ventilation**
- **Shading**
- **Watering of plant beds**

Taking **temperature** control first, the ability to heat the greenhouse adequately in winter will be the main problem, but retaining close control during the growing seasons and ensuring that a maximum is not exceeded will also need attention. The example greenhouse is sited on the south coast of England which, though frost is frequent in winter, is generally mild due to the benefit of a maritime climate. The median minimum temperature is −5°C. Were the greenhouse in central Wales the median minimum temperature would be −10°C. Table 15.1 shows the size of heater required to maintain minimum temperatures for two greenhouses, either of which is assumed to be partially protected from gales.

Table 15.1 Greenhouse heating

Floor Area	4.3 m²		7.2 m²	
Outside Temp. Minimum	−5°C	−10°C	−5°C	−10°C
+2°C	1 kW	1.5 kW	1.5 kW	2.3 kW
+7°C	1.8 kW	2.2 kW	2.7 kW	3.3 kW
+13°C	2.6 kW	3.2 kW	3.9 kW	4.8 kW

Plant growth does not take place at low temperatures and the survival of temperate plants can only be guaranteed above 6°C. Tropical plants cannot be catered for in such a simple greenhouse. The expense of maintaining a higher minimum can be seen from the table.

The temperature sensors used are the semiconductor type described in Section 2.5. Using the difference in V_{be} for two transistors operated at a fixed ratio of collector currents, they are arranged to provide a high impedance constant current source giving one microamp per Kelvin. This obviates the need for any calibration and they run from a very simple supply, through a current to voltage converter, to the input multiplexer. The two sensors measuring free air tempeature in the greenhouse are connected in series, as shown in Fig. 15.5 input S_4, giving the minimum value. The other two sensors are separately connected to measure propagation and seed bed temperatures respectively.

The temperature control is provided by two separate heaters. The small sixty-watt heater is formed from resistive wire zig-zagged across the base of the propagating bed and set in cement. This is quite adequate to raise the temperature from the chosen ambient temperature of the rest of the greenhouse. A two to three kilowatt domestic unit is all that is needed to provide the main greenhouse heating. A separately controlled fan is used to assist dispersal and reduce local hot-spots. Figure 15.4 shows the complete output interface. The output for the main heater is the most complex as it has the largest load. A solid state AC relay with isolation and zero-volt switching, is driven by a buffer amplifier from a PIA output to control the heater. Either the same circuit, which is described in Chapter 9, or the rather simpler direct-drive to a thyristor also shown in Fig. 15.4, can be used for the smaller loads.

Of course, the cost of heating the greenhouse could be cut if during the hours of darkness it could be double glazed. About 80% of the heat loss is through the glass with around 12% lost by air leakage and draughts and 8% through the floor. A recent innovation is the use of concertina-shaped air-filled bags which are deflated and lie flat on the floor during the day. They are inflated at night providing considerable insulation. These could easily be included in the design.

Turning to the control of **humidity**, solid state sensors have again been chosen for their simplicity in use. The sensors are formed by an exposed conductive layer over a non-conductive dielectric substrate. Changes in the humidity (from 0 to 100% relative humidity) cause the surface resistivity to change, the damp air acting as a variable resistor in parallel with the device. This type of sensor can operate over the temperature range −40° to +90°C which is ample for this application! Its response time is most interesting as in still air it can take tens of seconds for a 60% change in RH but with only moderate circulation this drops to the millisecond range. The accuracy of 0.5% is quite adequate. The sensors are shown used in the feedback path of operational amplifiers to increase sensitivity, though, in this case, a simple potential divider configuration could be acceptable.

Small solenoid valves are used to let water through from the main tank to the mist spray,

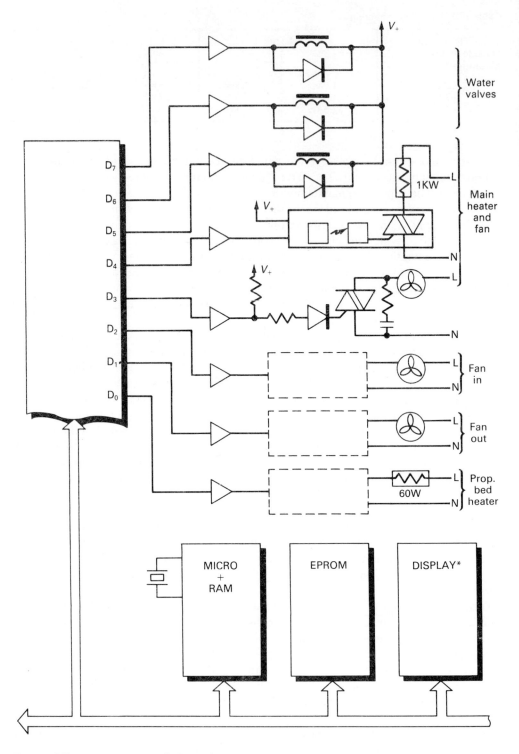

Fig. 15.4 Microprocessor controlled greenhouse output interface

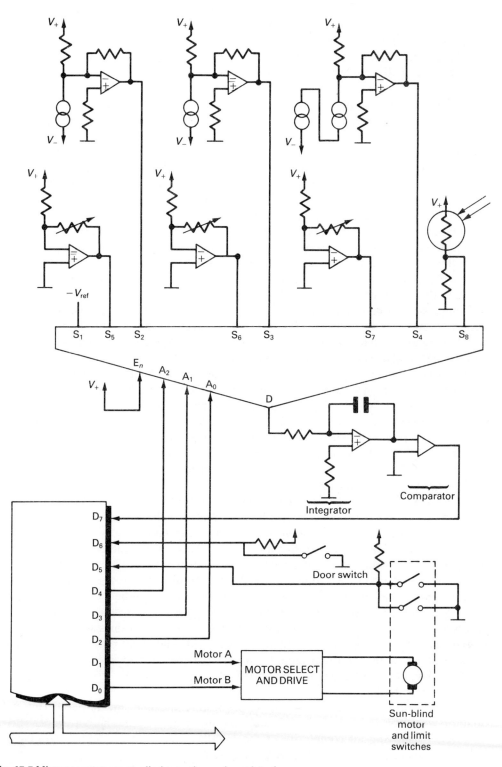

Fig. 15.5 Microprocessor controlled greenhouse input interface

to water the foliage and raise the humidity of the air, and to the water outlets into the beds to supply the root irrigation. The PIA outputs feed via buffer drivers to the solenoid coils in the way described in Chapter 9. As speed is not of the essence only the protective diodes are necessary. A relative humidity of 40 to 60% is maintained to ensure that mildew is kept at bay.

Ventilation will be required particularly in summer as in a warm, stagnant atmosphere fungus diseases flourish and young plants wilt. In hot humid weather, the air in the greenhouse will need to be changed up to twenty times per hour. Ventilation can also be used to assist temperature control by admitting cooler air and extracting warmer air. Two fans are used: one to exhaust air and one to take in fresh. They are about 25 cm in diameter and have louvres which close by gravity when the fans are not in use. The main heater fan can also be used to assist the movement of air. The fans can be turned on above about 12°C in summer and if the temperature rises more than 5°C above the chosen minimum in winter. The fans are controlled by similar solid state AC relay circuits to the heaters.

The amount of light or **shade** our plants receive can also be controlled—by a roller blind. This is located outside the glass so that it can assist in temperature control if desired—artistic licence places it internally in Fig. 15.3. The light intensity is measured by a simple potential divider with one resistor being a photoresistor. This is fed to one input of the multiplexer. The blind is controlled by a motor selected to drive in the correct direction by the two PIA outputs shown in Fig. 15.5. Its position is set by driving for known times, though micro-switches are included to give the zero and limit positions. A micro-switch is also fixed by the door so that an alarm can warn if it is left open.

The seven analog inputs (3 temperature, 3 humidity and 1 light) need to be converted to digital form. As the maximum frequency of any of them is less than 1 Hz, no sample/hold will be required and a dual-slope integrating conversion algorithm will prove adequate, as will eight bits of accuracy. By combining the switch between the unknown voltage input and the fixed V_{ref} reference voltage with a multiplexer an elegant solution is achieved. The analog multiplexer, for example a DG508 or a CMOS 4051, selects one analog input to connect through to the output when given a three-bit address. The three bits plus an overall enable input are decoded to eight separate logic signals which operate the field effect transistor switches. One FET is switched to give an ON resistance of a hundred ohms or so whilst the other seven are switched to give an extremely high OFF resistance. One input is thus selected through to the output of the multiplexer and then to the integrator and comparator. The dual-slope algorithm thus operates between one chosen input and input S_1 which provides the reference voltage, its operation being described in Section 6.4.

The remaining hardware is straightforward. An eight-bit microprocessor has been chosen and, as only a small amount is needed, the read-write store can be included on the same chip. An EPROM is used to hold the program which is fairly obvious in operation. The inputs are sampled in turn and converted where necessary. The values are compared with stored rules and the appropriate outputs are activated. The micro runs a counter to keep track of time and looks for user input to vary the parameters associated with the rules. A small display, keyboard and power supply complete the system but these are not described as they are standard items.

In a production unit the processor, both types of store, the timer and the parallel interface adaptor would be on a single chip leaving the minimum of external hardware. This is achieved by moving all complexity of the interface into software and integrating the interface and control aspects described in Chapter 1.

15.3 Aurac

The two previous designs are static and could be programmed in, for example, Pascal, Assembler or BASIC. Aurac is a mobile automaton and is an example of an extendable system and is programmed in an extendable language. It also has the special problem of mobility as it must **carry** its own compiler, interpreter and user interface for any commands or changes to be made.

Aurac's hardware is essentially very simple consisting of a 6502 microprocessor and, in the 1980 version described, only 1k bytes of random access store and 3k bytes of read-only store. Two parallel interface adaptors were used. One connects a 24-key keyboard, an eight digit (seven segment) display and cassette tape input and output drives. The second PIA provides all the real-world sensors, eight bits for input and eight bits for output as described further on. Figures 15.6 and 15.7 show Aurac with the display and keyboard visible on top and the schematic circuit respectively. The keyboard permits a 64 character set using shift keys and the seven segment displays give letters and punctuation as well as the numeric digits. A couple of the letters are difficult to fit on such a simple display and so have different patterns but the display is adequate for the purpose and characters scroll in from the right and off to the left. As can be seen Aurac is cylindrical, providing a base for load carrying, yet is able to rotate on its own axis and move in any direction. There are two balance wheels, at the front and rear, and drive motors located on each side. The motors are simple DC motors driving the wheels with foamed-rubber tyres through a gear train. This arrangement was

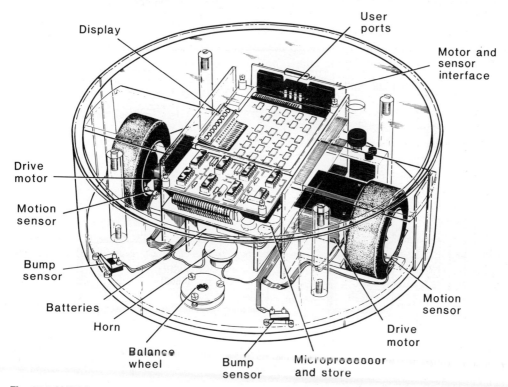

Fig. 15.6 AURAC

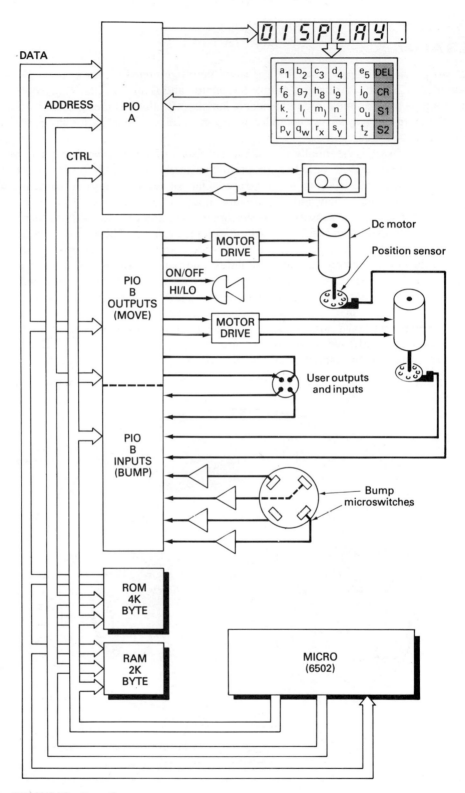

Fig. 15.7 AURAC schematic

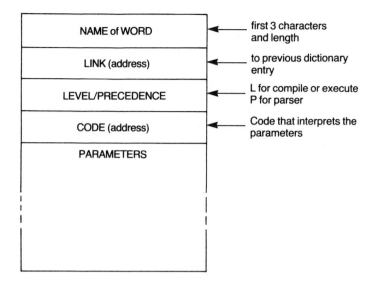

Fig. 15.8 AURAC word header

chosen to minimize current drain from the batteries. To provide feedback and positional information, motion sensors are included in each wheel using an infra-red source and sensor shining through a clear and opaque pattern in each wheel boss. The bump sensors are also shown and are triggered by movement of the outer perspex cylinder relative to the main part.

The detailed interaction necessary between the hardware and software of this example requires a simple yet powerful language. The language used in Aurac is derived from the work in the University of London on Snibbol, and by Charles Moore on Forth. There are four basic concepts involved. The first two are the notion of a **word** and a **dictionary**. The entire system in Aurac is structured into words held in a dictionary. These words correspond with procedures or subroutines of other languages. Whereas traditional languages only really have one type of procedure consisting of executable code, this family of languages (the **threaded interpretive languages**) have a major difference. The header of each word in the dictionary defines how it is executed, that is, whether the word contains machine code for direct execution, threaded code to be interpreted as links to other threaded or directly executable words, or just the value of a variable. Each word has a header containing the information shown in Fig. 15.8. The name of the word is stored in some abbreviated form, which for the 1980 Aurac was only the first three characters and the length. The next entries are only required by the high-level interpreter and are not required for direct execution. And, of course, some words contain pure code and hence require no parameter space in the body, using only the stack.

The user interface to the system is through a **high-level interpreter** which simply reads words from the keyboard, looks them up in the dictionary and then calls a shunting routine if required to convert prefix or infix notation to postfix (reverse polish). It then executes the words. This allows for user/defined data structures as in Pascal and similar languages but also for user-defined control structures, which is very convenient. Words may be defined permanently by using the inbuilt **compiler** to enter new words to the dictionary using the word **TO**. When initially interpreted at high level, operation is not fast, but once compiled

Table 15.2 AURAC inbuilt words

if	then	else	repeat	until	var
entry	compile	to	end	update	do
fence	shunt	shuntlitrl			( )

Threaded (High Level) Words

	Top		Top	Shunt_temp1
Dictionary {	Head	Tempdict {	Head	Shunt_temp2
	Limit		Limit	State

Inbuilt Variables

And	Or	Not	+	–	=
<	–> (assign)		@ (dereference)		Getword
Execute	Error	Scroll	$To	$Next	Findword
Number	Literal	Address	Value	Branch	CBranch
Top	Drop	Swap	ShPush/PoP		OpPush/PoP

Direct Machine Code Words

and using only the threaded linkage the code is surprisingly quick. The compiled code is significantly smaller than the equivalent assembler code and yet faster than most other compiled languages.

The words are obviously important and the majority of the **base** words will be the same for any system. A list of the words in Aurac is given in Table 15.2. As might be expected many of them are common to other languages as reserved words. There are only two words specific to this application for input/output. To make Aurac move, or perform any user-defined output or even sound its horn the word **move** is used. Move simply takes an eight-bit value as a parameter from the stack and outputs it to the PIA, thus turning the port and starboard motors, horn and user inputs, on or off. It does, however, check to see if the motors are running and permits them to move to the end of an increment, as measured by the motion sensors. If the move word is executed again before the increment is complete then the motor(s) are kept running to give smooth motion otherwise they (or it) are turned off at the end of the increment. This maintains positional accuracy as well as smooth motion in any direction. The bit pattern of the parameter for the move word is shown below:

User				
Outputs	**Hooter**	**Starboard Motor**	**Port Motor**
<1><2>	<ON/OFF><LOW/HIGH>	<BACK><FORWARD>	<BACK><FORWARD>

Thus to move forward the hexadecimal pattern given is <05>. To reverse a move <0A> is executed and to rotate clockwise a move <09> is given and so on. If a motor is selected with either <00> or <11> then it does not move.

To read in the sensors the word **bump** is issued. Bump returns the eight-bit value corresponding to the values of the eight input bits of the PIA. These in turn contain the states of the four bumper sensors, two user-defined inputs and the two wheel motion sensors. The pattern of the bits of the bump input is:

User		**Front**	**Rear**
Inputs	**Motion Sense**	**Bump Sense**	**Bump Sense**
<1><2>	<STBD><PORT>	<STBD><PORT>	<STBD><PORT>

Thus (BUMP AND 0F {hex}) gives just the motion sensors and so:

TO GO IF (BUMP AND 0F) = 0 THEN MOVE 5 END

defines a word GO which moves Aurac forward one increment unless the bump sensors indicate it has hit something, or has been kicked from behind! To make this more general a word can be defined to make any movement under these conditions:

TO GO IF (BUMP AND 0F) = 0 THEN MOVE ELSE DROP END

This word takes the parameter from the top of the stack and executes MOVE using this parameter, unless the sensors indicate otherwise when the parameter is dropped. To make a clockwise continuous rotation using the new word until a user input bit changes we define a further word;

TO CLOCKWISE REPEAT GO UNTIL (BUMP AND 80) = 0 END

When the word CLOCKWISE is input on its own then it is executed by the high-level interpreter and AURAC spins on its own axis until stopped by the user input <1> (hex pattern 80) or if one of the bump sensors is activated.

The importance of this approach to programming for interfacing and small control systems is emphasized by the fact that a number of chip makers have brought out versions of their single chip microprocessors with a Forth kernel held in on-chip read-only store. User programmed words are held in external ROMs. With a serial port, to permit user programming and control, counter/timer and parallel I/O lines included on the chip, a system can be up and running in a very short time.

Bibliography and References

Chapter 1   Introduction

Cripps, Martin. Inside The Computer or
Cripps, Martin. An Introduction to Computer Hardware.
Edward Arnold, 1977.
This level of understanding of computer hardware is assumed as a prerequisite for Computer Interfacing.

Bennett S. and Linkens D. A. Computer Control of Industrial Processes.
IEE Control Engineering Series 21, 1982.

Brignall J. and Rhodes R. Laboratory Online Computing.
International Text Book Co, 1975.
A balanced mix of mathematical techniques with rather dated hardware and software **relevant** to laboratory automation.

Mead C. and Conway L. Introduction to VLSI Systems.
Addison Wesley, 1980.
The classic (and excellent) work on VLSI systems using NMOS technology.

Weste N. and Eshraghian K. Principles of CMOS VLSI Design.
Addison Wesley, 1985.
An introduction to CMOS VLSI design directly following the pattern of Mead and Conway in approach and presentation which complements their book.

Chapter 2   Transducers

Allan R. New Applications for Silicon Sensors.
Electronics, **53** p 113, 1980.

Haslam J. A., Summers G. R., Williams D. Engineering Instrumentation and Control.
Edward Arnold, 1981.
Good Technician Engineering Council level coverage of transducers and measurement of common industrial variables for traditional control but with no interfacing or microsystems.

Electronics and Power.
IEE Publications, **28**, 1982.
An issue with special coverage of modern transducer techniques for many common variables.

Mylroi M. G. and Calvert G. (Eds). Measurement and Instrumentation for Control.
IEE Control Engineering Series 26, 1984.

Sahm W.H. Optoelectronics Manual.
General Electric Company, USA, 1976 et seq.
A general manual on optoelectronics but including optical isolation and some optical measurement techniques.

Sieppel R. G. Transducers, Sensors and Detectors.
Reston Publishing Co. Inc., 1983.
Clear description of common industrial transducers and their principles of operation.

Thompson J. Instrument Transducers.
Journal of Scientific Instrumentation, **34**, p 217, 1957.

Chapter 3   Sampling

Lathi B. P. Random Signals and Communication Theory.
Intertext, 1968.
An introduction to the mathematics of signal analysis.

Shannon C. E. Communication in the Presence of Noise.
Proc IRE, **37**, p 10, 1949.
The classic paper on this topic.

Slepian D. (Ed.). Key papers in the Development of Information Technology.
IEEE Press, 1973.
Includes Nyquist's original 1924 paper among a wide ranging selection.

Chapter 4   Conditioning

Bogner R. E. and Constantides A. G. An Introduction to Digital Filtering.
Wiley, 1975.
Thorough introduction to the theory and design of algorithms for digital filters, not just because they were at Imperial College.

Bozic S. M. Digital and Kalman Filtering.
Edward Arnold 1979

Garrett P.H. Analog Systems for Microprocessors and Minicomputers.
Reston Publishing, 1978.
Fairly wide coverage of analog filters and using operational amplifiers.

Gold B. and Rader C. Digital Processing of Signals.
McGraw Hill, 1969.

Chapters 5 and 6   Conversion

Clayton G. B. Data Converters.
Macmillan, 1982.

Dooley D. J. (Ed). Data Conversion Integrated Circuits.
IEEE Press, 1979.

Sheingold D. H. (Ed). Analog to Digital Conversion Handbook, 3rd edition.
Prentice Hall, 1986.
Derived from a manufacturer's data sheets and application notes handbook and now published as an informative book.

Tewksbury S. K. et al. Terminology related to the performance of S/H, A/D and D/A circuits.
IEEE trans. *Circuits and Systems*, **CAS-25**, 1978.

Chapters 7 and 8   Interfacing and Standard Interfaces

Bursky D. The S-100 Bus Handbook.
John Wiley and Sons, 1980.
A description of S-100 before it was 'standardised' as IEEE696, but fails to point out some of its serious drawbacks.

Colloms M. Computer Controlled Testing and Instrumentation – An Introduction to the IEC625 IEEE488 Bus.
Pentech Press, 1983.
A thorough introduction to using and designing for the IEEE488 with some examples.

Folts H. C. (Ed). Data Communication Standards – A Compilation.
McGraw Hill, 1982.
An extensive compilation of all **relevant** standards for data communication.

Kearsey B. N. and Jones W. T. International Standards in Telecommunication and Information Processing.
IEE Electronics and Power, p 643, 1985.
A detailed probe into the jumbled world of international standards making bodies which shows how their intercommunication does not bode well for our future standards.

Seyer M. D. RS232 Made Easy.
Prentice Hall, 1984.
A whole book on how to connect a plug to a socket. It is very clear and easy to read and illustrates (literally) many of the problems with the standard interface.

Chapter 9   Output Transducers

Acarnley P. P. Stepping Motors – A guide to modern theory and practice, 2nd edition.
IEE Control Engineering Series 19, 1984.
Examines aspects of stepping motor performance and capability for their application in industry.

Kenjo T. and Nagamori S. Permanent Magnet and Brushless DC motors.
Clarendon Press, Electrical Engineering Series 18, 1985.
Describes motor structure, drive and control circuitry with information on how to select suitable motors of these types for various applications.

Chapter 10   Environmental Constraints

Bennett W. R. Electrical Noise.
McGraw Hill, 1960.

Catt I., Walton D. S. and Davidson M. Digital Hardware Design.
Macmillan, 1979.
A most informative description of the problems of implementing high speed digital systems with their theoretical and practical solutions.

Denny H. Grounding for the Control of EMI.
D. W. Consultants Inc, 1983.

Morrison R. Grounding and Shielding Techniques in Instrumentation.
John Wiley and Sons, 1967.
Covers the basic considerations of electrostatics to show how to arrange optimum grounding and shielding for enclosures.

Ott H. Noise Reduction Techniques in Electronic Systems.
Wiley, 1976.
A very good coverage to assist in the design of microprocessor systems for use in electrically noisy environments.

Young R. E. Control in Hazardous Environments.
IEE Control Engineering Series 17, 1982.
Discusses the dangers of modern complex industrial installations and shows how to obtain more secure operation for oil, chemical and nuclear plants.

Chapters 12 and 13   Microprocessor Systems

Andrews M. Principles of Firmware Engineering in Microprogram Control.
Pitman, 1981.
A well rounded **description** of the design and implementation of processors which have regular microprogram control rather than being hard-wired.

Hanna F. K. (Ed). Advanced Techniques for Microprocessor Systems.
Peter Perigrinus Ltd.

Bibbero R. J. and Stern D. Microprocessor Systems, Interfacing and Application.
John Wiley and Sons Ltd, 1982.

Chapter 14   Design Techniques and Support Systems

Deplege P. (Ed). Software Engineering for Microprocessor Systems.
IEE Digital Electronics and Computing Series 3, 1984.
Thorough introduction to the techniques of software engineering which can be applied to microprocessor based systems and the differences with traditional software engineering on larger machines.

Rafiquzzaman M. Microprocessors and Microcomputer Development Systems.
Harper and Row, 1984.
Mainly reference material on the 8085, 8086, Z80, Z8000, 6809, 68000 and I432 processors, but it is worth reading the last three chapters on commercial microprocessor development systems.

Williams G. B. Trouble Shooting on Microprocessor Based Systems.
Pergammon Press, 1984.
Shows the common techniques used to find faults in micro systems and points to the design of systems which are easier to test.

Chapter 15   Design Examples

Brodie L. Starting Forth.
Prentice Hall, 1981.
An introduction to Forth, a threaded interpretive language, which shows by simple applications the power of this style of programming.

Pledge J. G. Microprocessor Controlled Gas Plasma Display.
MSc dissertation, Dept of Computing, Imperial College, 1981.

Sowden C. The Automatic User-Reprogrammable Ambulatory Computer.
Final year dissertation, Dept of Computing, Imperial College, 1980.

Index

Capital letters are used to denote mnemonic abbreviations.